Huynh Thi Phuong Linh

State-Society Interaction in Vietnam

ZEF Development Studies

edited by

Solvay Gerke and Hans-Dieter Evers

Volume 29

Center for Development Research (ZEF)
University of Bonn
www.zef.de

LIT

Huynh Thi Phuong Linh

State-Society Interaction in Vietnam

The Everyday Dialogue of Local Irrigation Management in the Mekong Delta

LIT

Cover image: PHONG NGOC HAI TRIEU – 2012

This publication is supported by BMBF – WISDOM PROJECT

This dissertation has been accepted by the Philosophical Faculty of the University of Bonn.

Bibliographic information published by the Deutsche Nationalbibliothek
The Deutsche Nationalbibliothek lists this publication in the Deutsche Nationalbibliografie; detailed bibliographic data are available on the Internet at http://dnb.d-nb.de.

ISBN 978-3-643-90719-6
Zugl.: Bonn, Univ., Diss., 2015

A catalogue record for this book is available from the British Library

Klosbachstr. 107
CH-8032 Zürich
Tel. +41 (0) 44-251 75 05 Fax +41 (0) 44-251 75 06
E-Mail: zuerich@lit-verlag.ch http://www.lit-verlag.ch

Distribution:

In the UK: Global Book Marketing, e-mail: mo@centralbooks.com
In North America: International Specialized Book Services, e-mail: orders@isbs.com
In Germany: LIT Verlag Fresnostr. 2, D-48159 Münster
Tel. +49 (0) 2 51-620 32 22, Fax +49 (0) 2 51-922 60 99, e-mail: vertrieb@lit-verlag.de

In Austria: Medienlogistik Pichler-ÖBZ, e-mail: mlo@medien-logistik.at
e-books are available at www.litwebshop.de

To Dad and In Memory of Mom

Preface

This book is dealing with state-society interaction in the field of local irrigation management in the Mekong Delta and provides new, detailed and very relevant insights into state-society bargaining processes in irrigation management. The study was written within the context of the BMBF financed WISDOM project, which aimed to establish a water information system for the sustainable development of the Mekong Delta in Vietnam. While fellow researchers were looking at knowledge production and knowledge sharing in the water sector, water policies, and the social implications of dam construction, Linh Huynhs' research was dealing with the question to what extent the state influences local practices in irrigation management and vice versa.

The water sector of the Mekong Delta is extremely complex and research in the Delta shows that water is economically and socio-culturally omnipresent, with a central role for the population and its diverse livelihoods. Recent developments in the delta region are directly linked to the management of the water resources: Flood control measures ensured successful rice intensification, fostered the aquaculture boom, enabled transportation networks and went along with urbanization and industrialization.

The book is based on a ten months intensive field research in Can Tho City province. It is investigating the state-society interactions and negotiations at the interface between actors in the bureaucracy and at the village level to provide a comprehensive explanation of how the state and other actors co-evolve in the process of learning from each other.

Looking at the role of the state and its interaction and negotiation with agents on the local commune and hamlet level, Linh Huynh's focus is on the power and influence of institutions at the local level and how these institutions, on the other hand, are shaped and influenced by local agents. The role of the state and its use and execution of power is exemplified in looking at canal dredging and field drainage, and the institutional bricolage reveals what exactly the authoritarian regime of Vietnam looks like. Vietnam has a history of local resistance towards policy implementation and Linh Huynh provides evidence that the state allows deviation from existing rules and thus releases power if the process is not undermining their exclusive political authority. Although considered an authoritarian state with a top down hierarchical structure, the Vietnamese state, i.e. its institutions, are able and willing to negotiate rules and accept resistance to policy decisions.

Looking at the impact of decentralization on irrigation management it is shown how decentralization affects the management of the irrigation sector and it reveals that the official decentralization is paralleled by an "unofficial" one which is very often guiding the execution and implementation of policies on the local level. The question, how policy is interpreted and finally implemented involves various stakeholders on the local level, including state representatives on district, commune and hamlet level, and it is the result of this process which then leads to action.

The author provides a very interesting and insightful overview of Vietnam's hierarchical administrative system of state agencies from the central to the local level in the light of the analysis of her empirical data. The study reveals the problems of having to share responsibilities in water resource management between different ministries at the central level, following this process down to commune level, where state representatives, who are very often farmers themselves, execute negotiation powers. It also shows the confusion about task division between agencies about which office takes responsibility for what. For the author the persisting question is whether this is a weakness of the state to achieve full control over the groups, or if groups find strategic ground at the interface between state's formality and community's informality. Huynh"s conclusion is that the dense structure of agencies from the central down to the hamlet level actually is a top down system of policy and mandates, but it is by no means inflexible. She is talking about a formal and an informal process, in which hierarchical state management structures co-exist with local informal structures on commune and hamlet level allowing a flexible approach in persuing agricultural tasks.

When it comes to social cooperation in irrigation management at the local level, special attention is on the negotiation process which includes the how, when, in what order etc. management of sowing, drainage, pumping and other activities in the agricultural cycle, which includes access to machines for canal dredging. The author proves that the - very often - day to day management of field drainage and canal dredging leads to a process of interaction between state and local actors in which processes are negotiated and choices are decided. These decisions are based on cost benefit analysis, but also social norms and social relations with neighbours and family. Very often state representatives are involved in the whole process as farmers, neighbours or family and solutions have to found to integrate old practices with new state induced rules and the creation of new formal/informal arrangements.

Huynh finally shows that state-society negotiations and bargaining processes reveal that best practice, although not totally in line with state regulations, will lead to changes in state policy towards agricultural activities at the local level. This process of cooperation is guided by the co-evolution of mutual learning. As long as the state authority is not threatened, the space for activities and flexible interpretation of orders is not only possible, by probably also wanted. The blurred boundaries between formal and informal activities in Vietnam can be found in many spaces and the author provides evidence of how the whole bargaining process of state, community and stakeholder negotiations looks like in detail. The state, Huynh concludes, is not weak, but shows its strength in not being the primary driver in a negotiation process, but a strong party in the bargaining process for favourable institutional conditions.

Solvay Gerke
Professor

Acknowledgement

This book and my own intellectual growth have only been possible due to support from many people and institutions. I take this opportunity to express my deepest gratitude to them.

I would like to express my deep gratitude to my PhD supervisor, Prof. Dr. Solvay Gerke, and my two tutors, Dr. Gabi Waibel and Dr. Hart Feuer, for their patient and constructive guidance and mentorship, and for their continuing encouragement.

I would like to give special thanks to the staff of Can Tho University, especially Dr. Nguyen Duy Can, for their guidance and administrative support for my access to research sites during data collection in the Mekong Delta. I am particularly grateful for the assistance provided by Hai Trieu during my field work. My sincerest thanks are extended to all staff members of the ZEF Doctoral programme as well as other staff at ZEF for making my stay in Bonn easy and comfortable and ensuring that my work at ZEF was a wonderful experience.

I would like to express my very great appreciation for the Vietnam state agencies, mass organisations at different levels, irrigation and agriculture experts, state cadres and farmers for their acceptance and support of my data collection. I would also like to offer my special thanks to the German Federal Ministry of Education and Research (BMBF) for funding my research and PhD program under the WISDOM project.

My journey would not have been that great without the special friends: Gabi, Siwei, Panagiota, Chau, Dunja, Maria. Thanks to them, other colleagues at ZEF and colleagues in the WISDOM project for their company, support, fruitful discussions, feedback, and friendship. My thanks also go to my other friends in Bonn and in Can Tho.

My greatest appreciation and love go to my beloved family for their love and strength throughout my studies abroad.

Huynh Thi Phuong Linh

2016

Table of Contents

List of Figures

List of Pictures

List of Tables

List of Abbreviations

CCP	China Communist Party
CPC	Commune People's Committee
DARD	Department of Agriculture and Rural Development
DPW	Department of Public Works under French colonial goverment
GSO	General Statistics Office
IMC/IDMC	Irrigation (Development and) Management Company
IMT	Irrigation Management Transfer
IRRI	International Rice Research Institute
ISF	Irrigation Service Fees
MARD	Ministry of Agriculture and Rural Development
MONRE	Ministry of Natural Resources and Environment
PIM	Participatory Irrigation Management
SEDP	Socio-Economic Development Plan
SIWRP	Southern Institute for Water Resources Planning
WISDOM	Water-related Information System for the Sustainable Development of the Mekong Delta
WUA	Water User Association

Prologue

> Cases like the one just described are the rule, not the exception. They serve to illustrate that the formal order encoded in social-engineering designs inevitably leaves out elements that are essential to their actual functioning. If the factory were forced to operate only within the confines of the roles and functions specified in the simplified design, it would quickly grind to halt. Collectivized command economies virtually everywhere have limped along thanks to the often desperate improvisation of an informal economy wholly outside its schemata.
>
> James C. Scott, *Seeing Like a State*

Commune Truong Xuan B, Can Tho City, 2011–2012 winter–spring rice season (Picture Hai Trieu) - On an afternoon in November 2011, 28 farmers gathered at the hamlet's Information House to participate in a hamlet meeting. In the very formal space filled with a red-and-yellow flag, slogans and a picture of President Ho Chi Minh, the farmers in their most common clothes — some men even not wearing their T-shirts — sit and discuss when they will pump the water so that the sowing and the rice season can be started. The gathering combined 2 meetings: the constituency meeting (representatives of the commune People's Council informed the farmers about the council's agenda and current state policies in agriculture and social affairs) and the pumping meeting (farmers discussed a pumping and sowing plan in relation to the state production plan and local conditions). The event was a combination of a state, bureaucratic, political event and a supposedly local gathering to arrange practice. In the pumping discussion, there coexisted the formal bureaucratic style of imposing the state agenda and an informal discussion between power equal individuals.

The Vietnamese state today encompasses the co-existence of the state's social–engineering designs and local social arrangements. The state's designs do not have all-encompassing power but neither are they completely dictated by local desires, as summarised in the proverb: '*The king's rules give into the village norms*'. To what extent, then, is it an event formal or informal? To what extent is it possible to distinguish the two? To what degree does the state penetrate and influence local practice, thus defining the state's status? What processes occur, and how so, at the local level where the state and its cadres come into contact with other stakeholders?

CHAPTER 1

ANTHROPOLOGICAL RESEARCH ON THE LOCAL DYNAMICS OF IRRIGATION MANAGEMENT IN THE CONTEMPORARY MEKONG DELTA

At the national level, the Vietnamese state, its role and status have been important subjects of research among Vietnam scholars. Endless debates on the type of the state have spurred exploration of the actual relations that the state establishes within the territory it manages. While such macro-level analyses have described thoroughly the evolution and status of the Vietnamese state in its unitary form, insufficient attention has been paid to the actual processes happening at the interface where state structure comes into contact with local structures. A few commentators, described shortly hereafter, have deeply investigated these processes. Their work has demonstrated that, instead of contributing to a more precise view of the Vietnamese state, studies at the interface invariably have found that functional governance is as diverse as the interfaces themselves, defined by sector, region, history and local agency.

In domains such as journalism and diplomacy, it is difficult to avoid coarse definitions of a state or a governmental regime, such as authoritarian or democratic. These categories gloss over the diversity within a state structure and provide poor explanations of the stability or effectiveness of a regime. So far, Kerkvliet (2001a, 2003, 2005) has conducted the most detailed, multi-scale summaries and analyses of the Vietnamese state. His work on Vietnam provides justification for more embedded studies of chronically misunderstood states and lessons about the limitations of any comprehensive characterisation of a state.

Kerkvliet's typology presents 3 non-exclusive views of the Vietnamese political system: dominating state, mobilisational corporatism, and dialogue. First, the dominating state sets rules and programmes through the superior power of the Communist Party and allows little or no space for independent organisation (Womack 1992, Thayer 1992, cited in Hy V. Luong 2003; Porter 1993, cited in Kerkvliet 2001a). Second, in mobilisational corporatism, various organisations established and run by the state and the Party function to help the state '*mobilize people to support its programmes and policies, maintain channels of communication between authorities and each sector of society, and manage social and economic groups that otherwise might become unruly*' (Kerkvliet 2003:31). Examples of such Party-related organisations include the Farmer's Union, General Confederation of Labour, Women's Union, Ho Chi Minh Communist Youth Union and Chamber of

Commerce and Industry, which are all managed by the umbrella agency, the Fatherland Front. The third and the least discussed description of the Vietnamese state is the dialogue state, which '*incorporates communication of contentious ideas and preferences in ways that, in Vietnam, are often indirect and nonverbal*' (Kerkvliet 2003:31). Kerkvliet argues that negotiations among the state, citizens and social groups outside the strict parameters of state policy and administration shape Vietnam's economy and society. A dialogue state can also be described as engaging in everyday politics, or political influence exerted in a bottom-up process that is not strictly collective[1]. Everyday politics encompass '*the debates, conflicts, decisions, and cooperation among individuals, groups and organisations regarding the control, allocation, and use of resources and the values and ideas underlying these activities*'. (Kerkvliet 1990, cited by Mollinga 2008:11).

In addition to the various definitions including different levels and sectors summarised and described by Kerkvliet, many other labels have been assigned to the Vietnamese state's exercise of power since the launch of Doi Moi in the late 1980s[2]. At the macro level, Wescott (2003:35) concludes that, despite state reform to allow a better fit for policy implementation and the possible involvement of local state agencies, a small number of Party members controls power and uses ideology to direct the state in all major issues. In relations between the state and local authorities, Koh (2001:534) reveals the pluralistic, unwritten rules of political elites within an authoritarian legal framework; the existence of ward-level governments with their own 'tiny states' is a sign of the Parkinson's State in Vietnam, with weak state regulation of local governance. Agreeing with Koh's (2001) argument for a weak state in rural areas, Fforde (2008b) describes how local leaders ignored state prescriptions to form informal farmer groups in the 1990s. However, the author raises the point that, as these local groups and village leaders engage in positive adaptations, they can

[1] Kerkvliet (2005:22) distinguishs three broad types of politics: official, advocacy and everyday politics.

[2] 'Doi Moi or the reform in the 1980s. In 1975, the country was united under a Communist regime, and collectivisation was imposed in the newly reunited South. Facing collectivised stagnation, reforms moved towards a market economy. In response to local resistance after a period of low effectiveness in agriculture collectives, the Vietnamese state in 1981 issued Resolution 100 permitting a partial return to family-based farming. The 6th Communist Party Congress officially launched reform in 1986 and introduced Doi Moi, which promoted market-oriented economy during a transitional period. Subsequently, Resolution 10 in 1988 confirmed the de-collectivisation of agriculture. Since then, the farm household, not the cooperative, has been considered the basic unit of agricultural production.

see these political problems as political opportunities (Fforde 2011:180). In such way, the author also notes the variety of interpretations of the Vietnamese state, both the customisation of Communism to gaps between theory and reality and the state's lack of political power to force local people to observe unwanted regulations (ibid:166). These categories which emphasise competition between the state and other actors might serve in journalism or politics but are insufficient to capture or explain the current dynamics in the relations or everyday dialogue which are the target of this book. The approach taken in this book on water management in the Mekong Delta can be characterised as everyday dialogue with the state. This approach extends beyond local negotiation and bargaining to look at the mutual learning or co-evolution between the state and local actors which refines local governance both officially and unofficially.

The everyday dialogue approach is especially useful in the rapidly changing social, ecological and economic environment of Vietnam, which puts pressure on governance to adapt faster than official policy. Today's market economy, changing way of life and state attempts at decentralisation, socialisation and retreating complicate state management. The diversity and dynamics in state management stem from evolution over time driven by interactions with other agents in society. These dynamics are perhaps most apparent in regions and sectors that are and historically have been in high levels of flux — such as water resources in the Mekong Delta.

Officially, water resources are under state management. However, water resources management in the Delta region has grown more complex due to on-going changes in legal, administrative, ecological and social conditions. Remarkable changes have occurred in policy itself, specifically the legal framework for water since the 1990s (Nguyen Thi Phuong Loan 2010a). In addition to the state management and technical structure, researchers have reported the (re)emergence and expansion of different agents, such as private enterprises, non-state organisations and community-based organisations (Norlund et al. 2006, Wischermann and Nguyen Quang Vinh 2003, Hannah 2007).

Economic liberalisation and socialisation policy in the form of equitisation has promoted the expansion of private stakeholders in the economy, such as the increase of private hydraulics construction firms in the Mekong Delta since the 1980s (Evers and Benedikter 2009a). The bureaucratic and sectoral complexity of the water sector, of course, takes place within the context of social change in the Mekong Delta, where the growth in the market mentality and monetisation have become increasingly important dynamics in state–society relations. While the scope of this research includes some discussion of the political and economic context in which state–society interactions take place, I focus squarely on the interface between farmers and the lower strata of the agricultural and irrigation apparatus.

Understanding local incentives and constraints from an anthropological standpoint is a critical avenue for exploring the relative agency of farmers in the state-mediated arena of farming in the Mekong Delta. A starting point for this research arises from project colleagues' work in the Delta, which suggests that '*local government institutions are both part of the State management system and bound to local interests and conditions*' (Waibel 2010:15). Although the local level is increasingly recognised as an important scale of research, water policies and academic writings have typically focused on state management structure (Waibel et al. 2012:22). Therefore, the present anthropological research provides an empirical account of local dynamics which both decide practice and contribute to the understanding of the state and its role in the contemporary Mekong Delta. Among many aspects of water management, irrigation and its relation to agriculture are especially relevant to the analysis of local practice and the state's role. The local practices of field drainage before sowing for the main winter–spring rice season and of canal dredging in the individual farming system, which are monitored by the bureaucratic state structure and involved in the interaction between the state and local stakeholders at the policy level and between the stakeholders at the local level, provide an opportunity to explore the state's role in relation to other stakeholders and to practice.

The aims of this research are to investigate the nature of state–society interactions at the interface between bureaucratic and informal areas and to determine the extent to which local actors can express agency and the degree of accommodation state actors allow. While analysing the local level of irrigation management, this research redefines the status, nature and strategies of the Vietnamese state and strategies and how local agents receive, adapt to and negotiate state policies in practice. The process of negotiating state policies is strongly influenced by local social arrangements interrelated to natural conditions and physical situations. In exploring the interactions between actors at the interface, the research does not seek to create a label for Vietnamese state or to advance the discussion on its loss of sovereignty or strength. Rather, the negotiation of institutions involved in irrigation arrangements provides a comprehensive explanation of how the state and other actors co-evolve in a process of learning. This work contributes empirical evidences to the debate about the state in Vietnam and contemporary local everyday dialogue that explains the local dynamics which decide practice.

Given these objectives, this research seeks to answer 2 questions:

- What are the informally agreed-upon elements of the platform of mutual learning and co-evolution for the state and local farmers?
- How do practices of irrigation management respond to the everyday dialogue between farmers and state managers?

The following discussion is based on the results of 10-month-long field research in the Mekong Delta. The macro-level state management in the irrigation sector was explored through expert interviews and visits to various state agencies in 2 provinces and 1 city and visits to state agencies in and around Can Tho city. Ethnographic methodologies with the case study approach were applied in 3 communes in the Thoi Lai district, Can Tho (See Figure A-17 in Annex), to collect the data on the negotiation of field drainage and canal dredging. The interaction of state agencies and local actors, including the local cadres, are analysed at the bureaucratic–informal interface. Data are contextualised in the broader theme of the institutional process and co-evolution of the state and local actors during the application of different state power dimensions to dynamic local arrangements. Details on the research steps, selected methodologies and analytical framework are explained in the Annex of this book.

This work consists of 6 chapters that cover the relevant background and provide an empirical account of the management of 2 irrigation activities: field drainage and canal dredging. Chapter 2 presents a literature review and the theoretical framework of this research, focusing on how diverse institutional dynamics are melded to form the landscape of functional state–society relations. Chapter 3 outlines the local natural and physical conditions and social arrangements and institutions which the state must accommodate when considering and implementing policy. This chapter focuses on the contemporary institutions that have been formed and renegotiated over this history of interaction. Chapter 4 examines the potential for state adaptation by looking at the space created for other actors with agendas, interests and ways of managing the system. Chapter 5 looks at the interactions themselves, focusing on the bureaucratic–informal interfaces: the physical boundaries of commune, hamlet, and pumping zone and the symbolic space surrounding field drainage and canal dredging, in which the practice of the collective activities is strongly decided by individual behaviour and informed by moral, emotional, social and economic factors. Chapter 6 brings together lessons learned from existing state–society accommodations and potential governance models to explore how the state and farmers arrange fraught but effective measures for adapting state policy.

CHAPTER 2

LITERATURE REVIEW AND THEORETICAL FRAMEWORK: ANALYSING STATE–SOCIETY RELATIONS AT THE BUREAUCRATIC–INFORMAL INTERFACE

To understand the interaction between the state and local actors in local irrigation management, this research concentrates on the local dynamics involved in the interaction and negotiation processes at the local level (commune and hamlet). While local administrative units supposedly are the site where policies are implemented, dynamics beyond state-designed procedures determine practice. This process involves different agents, including state cadres at various levels, farmers and their various organisations, and other private actors. Aiming to determine the extent to which agency can be expressed and to which state actors permit accommodation in state–society interaction in the Delta, this research borrows the term 'interface' (Long 1989, Long 2001, Long and Liu 2009) to analyse 2 irrigation activities: field drainage and canal dredging. At the bureaucratic–informal interface, the interaction between stakeholders in negotiating and bargaining individual behaviour which decides the formation of current institutions is analysed. Framing the local irrigation management in the concept of institutional process[3] from the realm of critical institutionalism and in Göbel (2011) theory of the dimensions of state power, this research analyses the local dynamics in the formation of institutional bricolage from the state–society interactions at the local level of irrigation management in the contemporary Mekong Delta.

This chapter presents a review of the literature on the labels and debates about the Vietnamese state, especially its relations to society. The second important objective of this chapter is to formulate the theoretical framework for this research, focusing on how diverse institutional dynamics are melded to form the landscape of functional state–society relations. This chapter first reviews institutional theories and concepts, from which Cleaver's (2012) institutional bricolage metaphor is especially suitable to explain the local dynamics of the management of the researched activities. Second, various views on the Vietnamese state are introduced, and the theories and literature on the state's

---

[3] Institutional processes, as clarified in detail in this chapter and throughout this work, involve complex human interactions in which stakeholders consciously and unconsciously shape, design, negotiate and alter the human-constructed constraints and opportunities created by rules and norms.

relations with(in) society are reviewed. The analysis of functional state–society relations is connected to Scott's (1987) peasant resistance. Göbel's (2011) 3 dimensions of state power serve as the analytical framework for investigating interactions between the state and local actors.

Negotiating the institutions of local irrigation management

While belonging to a community and set of organisations, human beings are always under the influence of rules (either state regulations or established, agreed-upon local arrangements) and norms often taken for granted. Researchers have questioned how arrangements become rules and/or norms, whether new ones replace old ones and whether rules and/or norms have the same effect on the behaviour of all individuals. This section explores the nature and characteristics of institutions. The concept of institutions has gained significant attention in the scientific world because of its capacity to understand human behaviour. Since the 1990s, the mainstream institutionalism school led by Elinor Ostrom has become popular and applied widely in irrigation management. This school of thought holds that

> Institutions are human-constructed constraints or opportunities within which individual choices take place and which shape the consequences of their choices. (McGinnis 2011, cited in Cleaver 2012:11)

This school posits that rational choices drive human behaviour, which has formed the foundation for such concepts as Hardin's tragedy of the commons and the prisoners' dilemma (Ostrom 1990). According to this school of thought, individuals tend to purposefully act in their own interests and self-consciously craft institutions '*to change the structure of repetitive situations that they themselves face in an attempt to improve the outcomes that they achieve*' (Ostrom 2005, cited in Cleaver 2012:12). The belief in rational choices creates incentive for promoting designed institutions.

> *In this approach, people's rationality is seen as 'bounded', their strategizing being limited by their ability to obtain and process the necessary information, particularly about the trustworthiness of others. This gives a role for properly designed institutions in which the operation of rules provides individuals with an assurance that others will use the natural resource in agreed ways, or be sanctioned*'. (emphasise in the original; Cleaver 2012:12)

According to the mainstream institutionalism school, resources can be better managed by reforming policy, building capacity, and redesigning community-level institutions to provide incentives for co-operation (Varughese and Ostrom 2001, Heikkila et al. 2011, cited in Cleaver 2012). Policy-makers have favoured this view because of its potential to bridge neo-liberal economic ideas and the desirability of decentralised local management and ownership (Ostrom 1990).

However, scholars from critical institutionalism argue that achieving the expected outcomes by managing human behaviour through designed institutions is not easy because individuals can perceive and translate rational choice and its importance differently. Cleaver (2012:xiii) argues that although designed institutions have potential, the inherently social nature of institutional functioning hinders the effectiveness of the strategy. This school emphasises the complexity of institutions entwined in every social life, their evolving historical formation driven by social and economic changes over time, and the interplay between global and local factors, the traditional and the modern, and between formal and informal arrangements (Mosse 1997, Cleaver 2001, Lund 2006, all cited in Cleaver 2012:8-9). '*Rules, boundaries and processes are "fuzzy"; people's complex social identities and unequal power relationships shape resource management arrangements and outcomes*' (Cleaver 2012:8-9). Scholars from this school offer various definitions of institutions:

> 'Institutions are neither animate things nor agents –they only exist in relation to people's interactions with one another. It is people, through their behavior and social relationships, who animate institutions'. (ibid:43). They are 'not being designed or crafted; [they] are patched together, consciously and not consciously, from the social cultural and political resources available to people based on the logic of dynamic adaptation (Smith at el. 2001:42)' (ibid:15). 'Institutions arrangement shape the way people manage material natural resources and infrastructure which is then influence the social arrangement to manage'. (Cleaver 2012:20).

In resources management, institutions are shaped both deliberately in formal spaces (which are to varying degrees public and formal; e.g. committees, associations, user groups, burial societies) and less consciously in routinised daily interactions (embodied in kinship and social networks, relations of reciprocity and patronage and sets of norms and practices deeply rooted in the habits and routine of everyday life). Overall, the critical school argues that crafting or designing institutions which achieved planned outcomes is impossible because institutions are formed, changed and reformulated through dynamic processes which play out in very different forms in varying contexts and, therefore, elude design (Franks and Cleaver 2007).

To overcome the 'myth' of local governance as 'good', traditional arrangements that do not produce social equality, Cleaver (1999:603) suggests eliminating the belief in that socially embedded institutions are 'better' than formal/organisational ones as local practices might uphold and reproduce locally specific configurations of inequity and exclusion. Sharing this view,

Meinzen-Dick and Nkonya (2007:14), water scholars studying African and Asian countries, argue that the pluralism of water institutions increased because of the plural characteristics of different kinds of law: state, customary and religious. While recognising the importance of local rules and norms in guiding decisions, Meinzen-Dick and Nkonya (2007:23) intentionally do not romanticise customary systems because they can reflect unequal power relationships in local communities, frequently dominated by local elites. Unless local rules and norms are analysed in relation to a specific group of people in society, judgments of whether these local designs are good or bad is relative.

Within the school of critical institutionalism, Cleaver recommends a metaphor for understanding institutions: bricolage. As bricolage, institutions are formed from pre-existing resources in order to make similar but new forms of institutions to guide practice at certain times. This metaphor highlights the role of institutions in consciously and unconsciously forming and changing rules and norms which are kept, returned or changed. Throughout this book, institutional bricolage helps explain the arrangements for field drainage and canal dredging selected by state and local agents. The characteristics of institutional bricolage are clarified and explored in those empirical accounts. Institutions in local irrigation management in the Mekong Delta are proven to be a cocktail of modern and traditional, formal and informal constantly negotiated in the quest for legitimacy.

The state appears as a stakeholder involving in the interactions that make up the institutional process. Thus, analysing the negotiation process can reveal the state's roles and relation to other agents in society. The following section presents various views of the state in contemporary Vietnam and introduces the concept of everyday dialogue, which this research demonstrates is the conceptual foundation for the interaction of state and society. Finally, this section describes the analytical framework based on Göbel (2011) concept of power dimensions of which is used to discuss these interactions at the bureaucratic–informal interface.

Identifying the Vietnamese state's roles and status at the bureaucratic–informal interface

Labelling by category (i.e. authoritarian or democracy) is a common practice in analyses of states and governmental regimes, especially in international politics and journalism. For instance, the United States State Department's Bureau of Democracy, Human Rights and Labor's 2012 report on human rights in Vietnams opens:

> The Socialist Republic of Vietnam is an authoritarian state ruled by a single party, the Communist Party of Vietnam.

The spectre of the authoritarian Vietnamese state is raised frequently in assessments of the country's political system, both in scientific articles and published reports, especially those related to human right or democracy. What then does the authoritarian regime of Vietnam look like? Linz (2000) asserts that authoritarianism manifests differently according to context.

> [These] systems are those with limited, not responsible, political pluralism, without elaborate and guiding ideology, but with distinctive mentalities, without extensive or intensive political mobilization, except at some points in their development, and in which a leader or occasionally a small group exercises power, within formally ill-defined limits but quite predictable ones. (Linz 2000, cited in Wischermann 2013:7)

Binary categories do not explain the diversity within a regime that makes a state become a more complex entity specific to a certain country. The nature of a complex regime informs the history of interaction between rulers and other stakeholders and the current dynamics that decide both practice and the stability of a regime.

The political setting in Vietnam has transformed significantly over history. Adding to and replacing the Confucian political system which took the patriarchal family as its basic model and did not concede any political or civil rights to citizens (Reis 2013:12), the Marxism-Leninism adopted in Communist Vietnam gave legitimacy to the collective and the society over individuals (Porter 1993:7). The systemic ideology strictly implemented in the North and later the South subordinates human freedom to the (party-defined) interests of the collective (Brocheux 2007, cited in Reis 2013:15). In the period before Doi Moi, belief in Communism and fear of losing control, especially in the newly reunited Southern land, gave the party–state incentive to use coercion to achieve uniform control over the country. After reunification in the mid-1970s, the Vietnamese state applied a system of hierarchical administration that, like the state elsewhere, was formed by a heavy-handed process of social standardisation and simplification. That process established a well-controlled administration at the expense of other forms of organisation (Marr 2004, cited in Larsen 2011). Although the state maintains a certain level of control in contemporary Vietnam, the post-Doi Moi state regime has undergone endogenous change in its roles and status influenced by various international and local factors, including the market economy and interactions with local actors.

With a complex structure, the Vietnamese party–state has been assigned different labels, not limited to authoritarian state, weak state, dominating state, mobilisational corporatism, and dialogue state. Amongst those, the categorisation of authoritarian, controlling or dominating state, while part of the vocabulary, is becoming less popular among contemporary scholars. Whenever recorded, the state attempt of control is partly secured by the domination of the Party state cadres in the emergent capitalist class and in the expansion of private economy in Vietnam, a result of the emergence of a market economy since Doi Moi (Cheshier 2010:18, Gainsborough 2005). However, recent analyses of the Vietnamese state have concentrated on the influence of pluralism which grants local governance a role in defining the state management style.

The power of local governance in relation to the state has been discussed within the contexts of the weak state or the loss of sovereignty. In relations between the state and local authorities, Koh (2001:534) argues that the existence of ward-level governments with their own tiny states is a sign of the Parkinson's State in Vietnam, with weak state regulation of local governance. Agreeing with Koh's (2001) argument for a weak state, Fforde (2008b) describes how local leaders ignored state prescriptions to form informal farmer groups in the 1990s. These theories place the state and other actors in competition for power and create a perception of winners and losers in state–society interaction, which can be inaccurate and does not necessarily reflect the actual interactions between the state and other actors in society.

Kerkvliet (2001a, 2003, 2005) contributes to the debate by summarising the state's diverse various ways of managing the country views and suggests adding the dialogue state based on the notion of everyday politics, which had not yet been discussed. Although the state dominates through restricting any attempt to criticise the ruling Party, the state can be described as mobilisational corporatist because of its structure of state agencies present at all administrative levels, from the central to the commune and even the hamlet. The relatively new idea of the dialogue state emphasises the possibility of interaction and negotiation between the state and other local stakeholders in forming policies and practices. This concept finds support in the bottom-up reform of the 1980s in which the state, facing local resistance, abolished the collectivisation policy. This process from fence-breaking to the '*step by step legalisation of illegal practices*' (Heberer 2005:215) demonstrates a specific characteristic of the Vietnam regime of reform: minimising loss of face (Rama 2008). Rama (2008) discusses the dialogue state, though not explicitly, when describing how pressure from citizens forced a response from the state and Communist Party, proving that there can be negotiation between various components of the state and interests in society. Yet, Kerkvliet and other scholars still seek a better definition of the state in Vietnam that can explain the coexistence of all three forms.

These various views of the state agree upon several characteristics of the post-Doi Moi Vietnamese state's mode of management. Under the current market-oriented economic reforms, the state faces resources shortage when implementing the policies associated only with a dominating state or mobilising corporatism (Woodside 1979, Thrift and Forbes 1986, cited in Kerkvliet 2003). Thus, as Waibel and Benedikter (2013:4) assert, '*Legal reforms and economic liberalization provided the foundation for a growing organizational pluralism, whereby the predominant mono-organizational socialism, as described by Thayer (1995) in the beginning of the 1990s, has been partly transformed*'. In addition, despite the hierarchical administrative system reaching from the central government to the province, district and commune, the Vietnamese government has often resisted the authoritarian temptations of high modernism (see Scott 1987) because of the existence of and belief in a private sphere of activity; the impossibility of an economy sovereignty (Foucault) in which a far too complex economy makes it impossible to manage in detail by a hierarchical administration; and the existence of working, representative institutions through which a resistant or grassroots civil society can exert influence (Scott 1998:102). Explanations for the state transformation and retreat and deviation in policy implementation due to local responses vary from a weakened state's inability to impose control to the state's ability to adapt to and accommodation new circumstances.

Thus, another possible interpretation of the state is that it has been transformed but has not necessarily weakened. As Fforde (2011:180) asserts, the political leverage of local leaders in informal farmer groups can also be seen as either political opportunities or a kind of positive adaptation. Although some might criticise Vietnam's one-party system under criticism, others argue that context matters. For instance, the way that Vietnamese leaders insist on no need for opposition parties '*is not just a crude defence of authoritarianism but represent heartfelt opinion based on a very different view of State and opposition than that of the West*' (Gainsborough 2010:22). Consequently, political conditions in Vietnam can also be seen as the application of 'metagovernance' or 'governance of governance', as defined by Bell and Hindmoor (2009, cited in Keating 2010:104):

> Metagovernance encompasses the practices and procedures that secure governmental influence, command and control within governance regimes. Six key functions which they consider are the prime responsibility of the state: steering; effectiveness; resourcing; democracy; accountability; and legitimacy.

According to the argument that the '*top-down government is not dead*', the state's responses to pressure from citizens are strategies in new modes of governance to achieve its goals and objectives in a changing world with an increasingly individualistic society and less tolerance of authority (ibid: 104). Hibou (1999, cited in Gainsborough 2005) shares these view of the state's role:

> Contracts or agreements, whether formal or informal made between state and private actors are far from being permanent or even long lasting. On the contrary, they are deliberately unstable, even volatile and secret and up for negotiation all the time. This instability is not the result of poor management or other inadequacies. Nor is it the expression of external dependency. It is rather at the heart of politics: creating and maintaining conditions for the 'exercises of power' or 'mode of governing in themselves. (Hibou 1999, cited in Gainsborough 2005:25)

In this way approaching, the de-collectivisation of the 1980s, the enactment of grassroots democracy amid local unrest in the 1990s and the relative acceptance of local deviation in policy implementation can be translated as the strength of state in applying metagovernance. If using metagovernance (Bell and Hindmoor 2009), the state then can adapt to the new context. The state might release powers (allow deviation) but then perceive the need to 'retrench' if its exclusive political authority is undermined. Thus, terminology for Vietnamese state becomes a moving target, difficult to pin down as it co-evolves with the nation, citizens, and/or world pressures.

While affirming the state nature of negotiation, the dialogue concept also highlights that local actors play an important role in interactions. The response of local agents frequently has been conceptualised within the theme of peasant resistance. If individual, peasant resistance not only has no risk but also has certain advantages, such as flexibility and persistence which compensate for a lack of central capacity for coordination (Scott 1987:297). According to Scott, the sub-ordinate classes (farmers, in this case) are far less interested in changing the larger structures of the state and the law. Rather, they use ordinary weapons, such as '*foot dragging, dissimulation, desertion, false compliance, pilfering, feign ignorance slander, arson, sabotage and so on*' (ibid) because, the author contends, of the fear of repression or Marx's '*the dull compulsion of economic relation*' (ibid:246). While the former was relevant in the previous period when the state wielded sanctions, today, economic and technologic benefits to the state drive silent local adjustment, instead of open confrontation. No longer is in contemporary Southern Vietnam that:

> Lacking any realistic possibility, for the time being, or directly and collectively redressing their situation, the village poor have little choice but to adjust, as best as they can, to the circumstances they confront daily. (Scott 1985:246)

Today, farmers of the Delta have decision-making power, as is proven in this research. With individual farming and the new mode of state governance (mobilisation instead of coercion), practice is decided based on cost–benefit calculations after considering all factors, including state guidance and other local natural, physical and socio-economic conditions. Thus, one can call these actions resistance in covert and soft forms or simply adaptations or decisions made in negotiation. It can be said that not only farmers do adjust to state policies, but the state also adjusts its policies to demands and sometimes pressures from farmers.

While opening itself to negotiation, the state has fears in high modernism the loss of power and, thus, the legitimacy of the state. As Scott (1998:305) finds, the more the cultivator knows, the less important the specialists and their institutions – including state regulations and scientific knowledge. Therefore, since the late 1980s, the Vietnamese state has been the major actor in some rural economic sectors (land ownership, grain trade, and credit system), strongly influencing the negotiation and interaction with and between farmers, domestic and global markets. However, the state's role '*doesn't seem so intrusive and authoritarian as in the past*' (Diglio 1998:45). The question then is how the state today accommodates local responses while controlling or trying to control the process of bricolage passively and directly and how those methods support or do not support the various views.

This research investigating the nature of state–society interactions at the bureaucratic–informal interface is guided analytically by the direction of the dialogue state. The dialogue that explored in this research is specific to the everyday negotiation of institutional bricolage which is strongly dependent on the bargaining of individual behaviour. Thus, empirical evidences demonstrate the everyday dialogue in which the state and local actors involved, including local cadres, farmers and farmer organisations, co-evolve through the process of mutual learning to refine the local governance. The everyday dialogue which decides local irrigation practices is explained analytically by Göbel's (2011) framework of 3 dimensions of state power.

Göbel's framework of the dimensions of state power

This research adopts Göbel's approach to analyse the state (2011) as it manifests in 3 different dimensions of power: despotic, infrastructure and discursive power. According to Göbel (2011:186), authoritarian regimes differ from each other in the extent and scope to which they use the 3 power dimensions (see Figure 2-1).

Figure 2-1: Indicators of the 3 power dimensions of an authoritarian state (Göbel 2011:186)

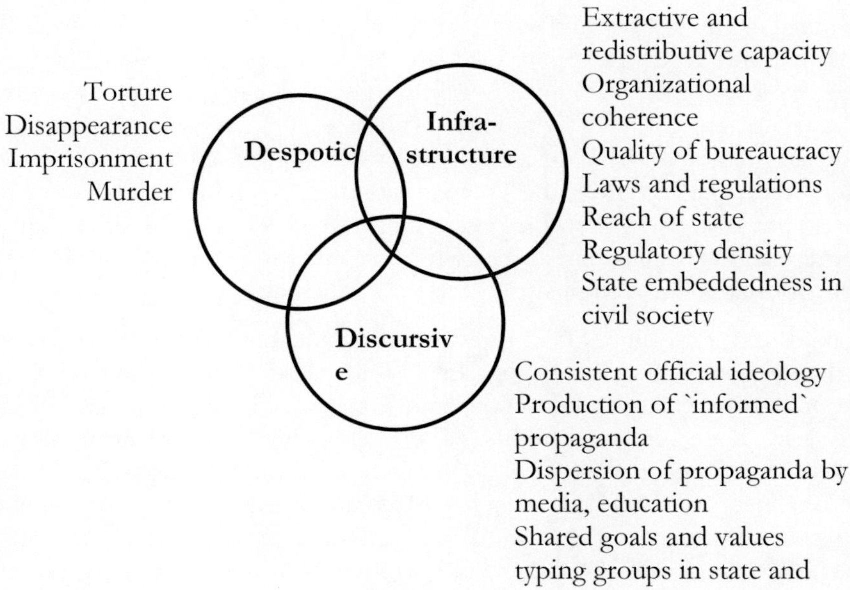

Göbel categorises a state based on its capacity to wield 3 kinds of power: the coercive power to impose its will on the people (despotic power; the regulatory power exerted by institutions and organisations (infrastructural power), including the state embeddedness in society; and the power to make people want what the government wants them to want (discursive power (ibid:177, Lukes 1975/2005). To exercise infrastructure power, the state places a '*dense network of institutions (infrastructure) to provide citizens with incentives to behave in certain ways and thereby reduces complexity and improves predictability*' (ibid:183). Through infrastructure (e.g. legal documents and administrative structure and physical infrastructure such as canals), the state implements policy and maintains a presence in even the remotest territories (Soifer and vom Hau 2008, cited by Göbel 2011:185). Additionally, the embeddedness of the state at most local levels enables it to react adequately to local demands, increasing its legitimacy (Mann 1984, ibid:185), and to undertake coercion activity when local tensions reaches threatening level. The state must find the most distant way of remaining embedded. Too much embeddedness costs resources; too little, and the state risks losing legitimacy.

The last and supposedly highest level of power, discursive power offers '*means of securing the active complicity of the subjects of power in their own self-regulation*' (Jessop

2008, cited in Göbel 2011:186). Discursive power includes both political propaganda, which shapes political and social realities, and the correspondence of the state to reality. Elites can impose social order by creating and disseminating discourses and concepts which define what is true, beautiful, moral, fair and legitimate (Scott 1987:38). Thus, '*they build a symbolic climate that prevents subordinate classes from thinking their way free*' (ibid). Discursive power gives the state has more low-cost options to address social problems and regime challenges than repressing or cracking down on opponents. As this is the highest level of state power, the Vietnamese state can, and does, focus or even increase propaganda and information control efforts and pays less attention to other domains. This research sheds light on whether the state achieves its goals or faces obstacles through this approach.

A state can exercise all 3 kinds of power, and the choice of which power type and the extent to which it is applied to a specific activity or event illuminates a facet of the state, though not enough to categorise or label the state. When using despotic power, the state becomes dominating. With the nature of state standardisation and simplification, infrastructure power is used at all times but to different extents. Through using infrastructure power, the state illustrates its mobilisational corporatist nature. The less-dominating mobilisational regime also effectively transfers local demand into decision making during the process of embeddedness. Lastly, by promoting mobilisation and the process of embeddedness, discursive power creates space for local involvement, which, with or without soft despotic power, might pave the way for a dialogue state. Thus, a state can be located at the juncture of all 3 circles and, depending on the sector and event, will exercise one, all or a combination of the 3 power types to different extents, creating evidences for those who want to give a certain label to a state regime.

The use of 3 dimensions of power is proven through the analysis of the making of water landscape of the Mekong Delta and the management of field drainage and canal dredging in which the state envisages its role in interactions and negotiation with other stakeholders. This ethnographic research on negotiation illustrates that, in local irrigation management, the present-day Vietnamese state mostly exercise infrastructure and discursive power. Through the embeddedness process and mobilisation, the state allows certain levels of deviation that do not undermine state power, and in return, local actors, including state cadres, consciously and unconsciously negotiate state policies and other local arrangements to maximise their advantages. In negotiating and refining local governance, local actors engage in an everyday dialogue with the state, facilitating co-evolution, instead of defensive resistance.

CHAPTER 3

THE MAKING OF A WATER–CONTROLLED LANDSCAPE - THE AGRICULTURE AND RURAL MEKONG DELTA

Chèo ghe sợ sấu cắn chưn

Xuống sông sợ đỉa, lên rừng cọp um

[Take the boat afraid of crocodiles biting the leg.

Go down to water afraid of leeches. Go up to the forest afraid of tigers.]

- Southern Vietnamese folk

Generations have narrated this folktale to express how the Delta looked like when the first Vietnamese settlers from Tonkin arrived. In the 18th and 19th centuries, the Delta possessed a diverse ecological landscape, where waterways, swamps and forests hid both risks and opportunities for human settlements. Surprisingly, about 2 centuries later, the Delta with '*sacred forest, poisonous water* [with many dangers of mosquitoes and crocodiles]' had become a human living-space with irrigated agriculture supported by hydraulic works, such as dykes, canals and gates, and today, the Delta supplies most of the nation's agricultural production. It has been transformed '*from a "water landscape' into a modern "hydraulic landscape", i.e. a "hydraulic society" in which hydraulic management plays a crucial role in many aspects of daily life and the economy as a whole*' (Evers and Benedikter 2009b:10). The making of the water-controlled landscape through various periods of destruction and (re)building has transformed not only nature-dependent, rain-fed agriculture into hydraulic, irrigated agriculture but has also shaped the Delta's contemporary, rural society. How that transformation has taken place over the pre-colonial, colonial, and reunification periods and how human investments and technologies have formed the present Delta and its dynamics surrounding water resources, agricultural production and rural life are at this chapter's centre of analysis.

This chapter presents an analysis of the making of the Delta's hydraulic water landscape, its specific irrigated agriculture and it rural life. First, this chapter gives an overview of the Mekong Delta, the importance of irrigation and the diverse landscape divided hydrologically into 6 hydro-ecological zones. Second, based on the history of hydraulic interventions, the making of the hydraulic Delta is shown to be guided by the interrelations among natural conditions, existing hydraulic infrastructure and social, cultural and political circumstances. The chapter explores the historical influence of natural and social conditions and political regimes on the *work-never-ends* water infrastructure interventions in the Mekong Delta. Human interventions in hydraulic infrastructure have affected and guided the transformation of agriculture and rural livelihoods into the hydraulic society of the Delta today, while the transformation in agriculture has also required continuity in building and rehabilitating infrastructure, allowing the state to play a major role in designing and realising interventions. Next, this chapter explains the specific characteristics of Delta society which have been formed and negotiated over history. I argue that the Delta life and water landscape have been formed through the continuous interactions of human and nature in which new interventions have been built on existing infrastructure closely linked to natural conditions, the socio-economic situation and the state policies. This historical case study of the Delta contributes to the school of critical institutionalism which views institutions as a process in which people adapt and negotiate various influencing factors too complex to be designed and subjected to continuous changes (Cleaver 2012).

Diverse landscapes of the Mekong Delta and the domination of hydraulic infrastructure

At the end of the Mekong basin[4] in Vietnamese territory lies the Mekong Delta (Figure 3-2), which is the most downstream and the biggest region in the larger Mekong Delta stretching into both Cambodia and Vietnam. In Vietnam, it is called the Western Region (Miền Tây) for its location in South Western of Vietnam, bounded by Cambodia and Ho Chi Minh City in the north, the East Sea in the Southeast and the West Sea in the west. The Delta covers 39,000 sq km (22% of Vietnam's territory) with about 600 km of coastline and is divided into 12 provinces (Long An, Tiền Giang, Bến Tre, Trà Vinh, Vĩnh Long, Đồng Tháp, An Giang, Kiên Giang, Hậu Giang, Sóc Trăng, Bạc Liêu and Cà Mau) and 1 central city, Cần Thơ. The Delta has approximately 17.4 million inhabitants in the Delta, constituting 12% of Vietnam's total population (data 2012, General Statistics Office - GSO 2012). The territory is shared by 31 ethnic groups, of which the Kinh account for 79%, the Khmer 15% and Chinese 3.9% (Southern Institute For Water Resources Planning - SIWRP 2011b:42).

[4] The Mekong basin covers 795,000–800,000 sq km, is regarded as the world's second-richest river basin in biodiversity and has a population of 60 million from more than 95 ethnic groups (World Wide Fund for Nature 2004, cited in Le Anh Tuan et al. 2007:18). The Mekong River travels 4,800 km from the Tibetan plateau in China downstream through Myanmar, Laos, Thailand and Cambodia until it empties into the South China Sea at the southern tip of Vietnam, where lies the Mekong Delta.

Figure 3-2: Provinces of the Mekong Delta and the Hau and Tien rivers

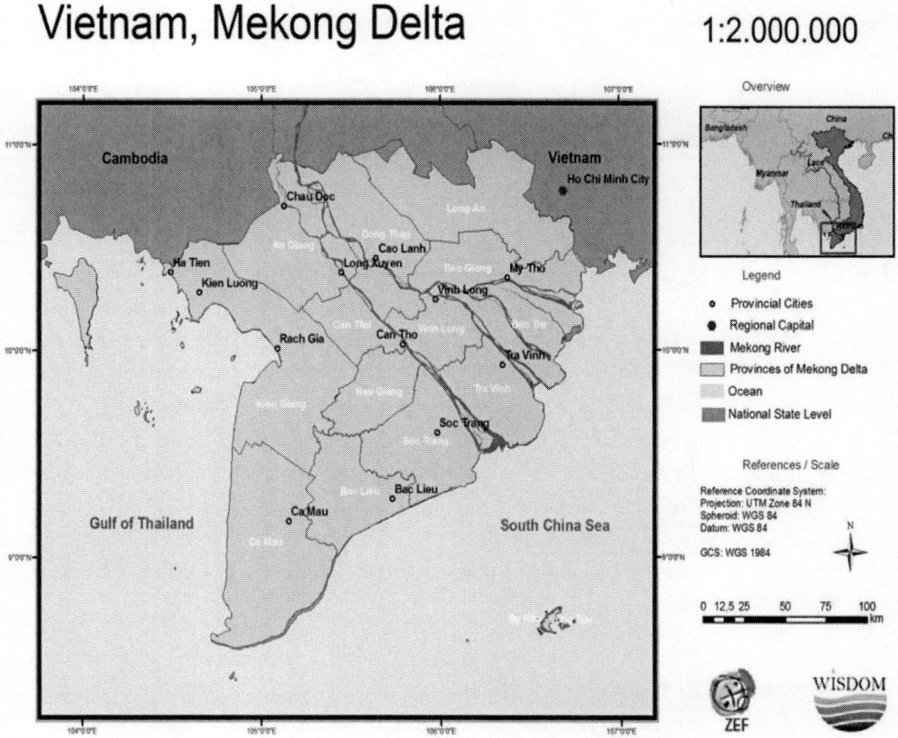

Source: ZEF-WISDOM

Physically, the Mekong Delta is classified as flat, with most elevations from 0.7 to 1.2 m (Southern Institute For Water Resources Planning - SIWRP 2011b:8). The Delta has:

- 1.2 million ha of alluvial soil along the Hau and Tien rivers favourable for agricultural cultivation
- 1.6 million ha of acid sulphate soil distributed in the Dong Thap Muoi (Plain of Reeds) and Long Xuyen Quadrangle
- 0.75 million ha of salty soil along the coastline
- 0.35 million ha of other soil types, including grey soil in the ancient alluvium, peat soil and mountainous areas (Le Sam 1996:8, Nguyen Van Sanh et al. 1998:18-19).

The Delta is characterised by a monsoon climate[5] with roughly 6 months of sun (December–April) and 6 months of rain (May–November). The second half of the rainy period and the tidal regime create up the flooding or water-rising season. The average annual rainfall is 1,700 mm (ranging from 1,200 mm to 2,400 mm), which is unevenly distributed, with most falling during the 6-month rainy season (Pham Cong Huu 2011). The Delta has 2,226–2,790 hours of sunlight a year and fewer natural catastrophes than the central and northern regions of Vietnam. The natural Delta provides fertile land and abundant water resources that support human settlements and agriculture but also certain obstacles for which people have to find the ways to avoid, adapt and/or control.

Hydrologically, the Delta's landscape consists of the dense water network of the Mekong River system, its tributaries and thousands of smaller canals. The Mekong River enters Vietnam in An Giang province, has 2 main branches — the Hau (Bassac) River and the Tien (Mekong) River — and travels to the sea through 9 estuaries: Tieu, Dai, Ba Lai, Ham Luong, Co Chien and Cung Hau from Tien River; and Dinh An, Ba Thac[6] and Tran De from the Hau river. In Vietnam, the Mekong River is named the Cuu Long river (River of Nine Dragons), leading to the Cuu Long Delta. Annually, the Mekong River has an annual flow volume of 366–448 billion m^3 (Bucx et al. 2010:23) and transports 150–200 million tonnes of sediment into the Mekong Delta (Pham Quang Tu et al. 2013:11).

With these geographical conditions, the Mekong Delta presents both opportunities and challenges to its inhabitants and their activities. The huge amount of water in the Delta has been put to multiple uses, and the sediment supplies nutrients for annual agricultural cultivation and in the diversity flora and fauna in the ecology (Duong Van Ni et al. 2001, cited in Fuhrmann 2008:14). The large, flat alluvial plain with abundant water resources (despite the uneven distribution of water flow) and annual siltation providing rich nutrients has made the Delta Vietnam's most productive agricultural region. It serves as the nation's rice granary and supplies the majority of national agricultural production, including orchards, vegetables and fishery products. According to the GSO (2012), the agriculture sector occupies 2.6 million ha of

---

[5] The seasonal distribution of dry and wet months depends on the monsoon season. The dry season usually coincides with the North-East monsoon with dry heat and little rain, while the wet season coincides with the South-West monsoon characterised by high temperatures, humidity and rainfall (Bucx et al. 2010:17).

[6] Historians have searched for signs of Ba Thac estuary (the third estuary of the Hau River) of the Cuu Long River but have found no explanation.

the Delta's nearly 4 million ha; the rest is devoted to forestry, special use land and housing (data 01.2012). In addition to these natural resources, the Delta poses such challenges as flooding (with both benefits and challenges), acid sulphate soil and salinity intrusion. With the cyclical flow of brackish and fresh water, for example, extended drainage can risk alum build-up and decreased fertility (Duong Van Ni et al. 2001, cited in Fuhrmann 2008:14). The Delta's rich sedimentation can hinder canal maintenance, particularly canal dredging (see chapters 4 and 5 for the analysis).

The opportunities and challenges of the Delta together drive the complexity of water management for both farmers and the state. Taking flooding as an example, the flooding season (*mùa lũ*) or water rising season (*mùa nước nổi*) takes place from July or August to around November or December when abundant amounts of water come from rain, upstream (from Cambodia), and estuaries pushed inland. Those natural effects, combined with social phenomena exacerbated by the tidal regime, past policies and alterations to terrain (Biggs et al. 2009:97), can create unexpected inundation in certain areas of the Delta. Le Anh Tuan et al. (2007) estimates that 1.2–1.9 million ha of the Delta flood annually (ibid:25), presenting both risks and benefits to Delta residents. While flooding causes death, production losses, bank erosion, navigation hazards, pest, toxic plant invasions, and health risks, the water rising season also provides water and fertile sediment to the fields, creates an environment for fish spawning, supplies a rich source of aquatic products, flushes toxins and farm pollutants such as agro-chemicals from the acid surface of soil areas, assists in pest control strategies and balances the water quality and ecology (Le Anh Tuan et al. 2007:25, Pham Cong Huu 2011:66-69). The dual nature of the Delta is one of the main influences on human interventions in the Delta, especially in the choice between control and adaptation, between using the Delta's resources and avoiding or controlling its challenges. The history of interventions elaborated in latter sections of this chapter shows the influence of these natural conditions.

Domination of hydraulic infrastructure in the Delta's water landscape

In the contemporary Delta, it is difficult to see the vast swampy areas with 'scared forest and poisonous water'. Instead, what can be seen are human settlements with irrigated agricultural land and a dense canal system supplying water for various purposes, many regulated by modern hydraulic structures not limited to dykes, sluices, culverts and pumping stations. Amongst the many uses of water resources in the Delta, irrigation is the most important to sustaining the agricultural production which is vital for farming households, national food security and the exports sector. Contemporary Delta's agriculture is far different than traditional rain-fed or flood-fed rice cultivation, which allowed 1 season per year. Traveling around the Delta on a motorbike on newly paved, 3–5 m rural roads, one can easily see rice plantation throughout the year and the diversification of farming activities. This is called irrigated agriculture or, in Benedikter's (2014) terminology, hydraulic agriculture. This agriculture supplies 38% of the Delta's gross domestic product (ibid:22). In 2008, farmers in this South Western Region provided 70% of fruits, 40% of fishery products and 74% of aquatic products in the nation (Southern Institute For Water Resources Planning - SIWRP 2011b:6).

To maintain and enhance important agricultural production while protecting the Delta from natural disasters such as flooding and salinity, the landscape has been subjected to central and local interventions of hydraulic infrastructure intended to provide transportation, flooding protection and mitigation, and year-round irrigation. Over history, water infrastructure has become a dominant subject of state and ruler interventions, recognised as necessary for the development of irrigated agriculture and solutions for the challenges of flooding, salinity and acid sulphate soils. These interventions have resulted to the construction and rehabilitation of canal networks and large-scale irrigation and/or drainage schemes (Benedikter 2014b).

Consequently, the contemporary Mekong Delta presents a dense network of rivers, canals and various hydraulic structures (Figure 3-3). The 2 main rivers are the Hau and Tien rivers, with an average width of 1,000–1,500 m and average depth of 10–20 m (up to 40 m in some sections). From the 2 main rivers, water is conveyed through a dense system of canals of different sizes classified by the state in 3 levels primary, secondary, and tertiary (state categorisation) Southern Institute For Water Resources Planning - SIWRP 2011b:10). In local terms, canals are divided into larger rivers (*sông*), smaller natural rivers and creeks (*rạch*), artificial canals (*kênh*) and irrigation ditches (*mương*) (Benedikter 2014b:23). The Delta has more than 15,000 km of main and primary canals, nearly 27,000 km of secondary canals, and approximately 50,000km of tertiary canal. The canal system has a density of 80–100 m per hectare of land, including sluices, culverts, dykes[7] (13,000 km, with 7,000 km of embankments for flood protection for the summer–autumn rice fields and more than 500 km of sea dikes), pumping stations and small pumps. This system forms the basis of the Delta's water infrastructure (Bucx et al. 2010:30). According to the SIWRP (2011:30), in 2011, the Delta's hydraulic works met

[7] In addition, more than 200 km of dikes to hold water for national park and mangrove forest production.

72 % of irrigation demand for existing arable land and up to 60% of demand water intake, salinity control and drainage of flood and brackish water.

The dense canal network and head-works have become the backbone of Delta's life and economic activities. Additionally, the human desire for settlement and existing bio-natural conditions continue to drive the preference for hydraulic infrastructure in the human engineering of the Delta's landscapes.

Figure 3-3: Water infrastructure of the Mekong Delta

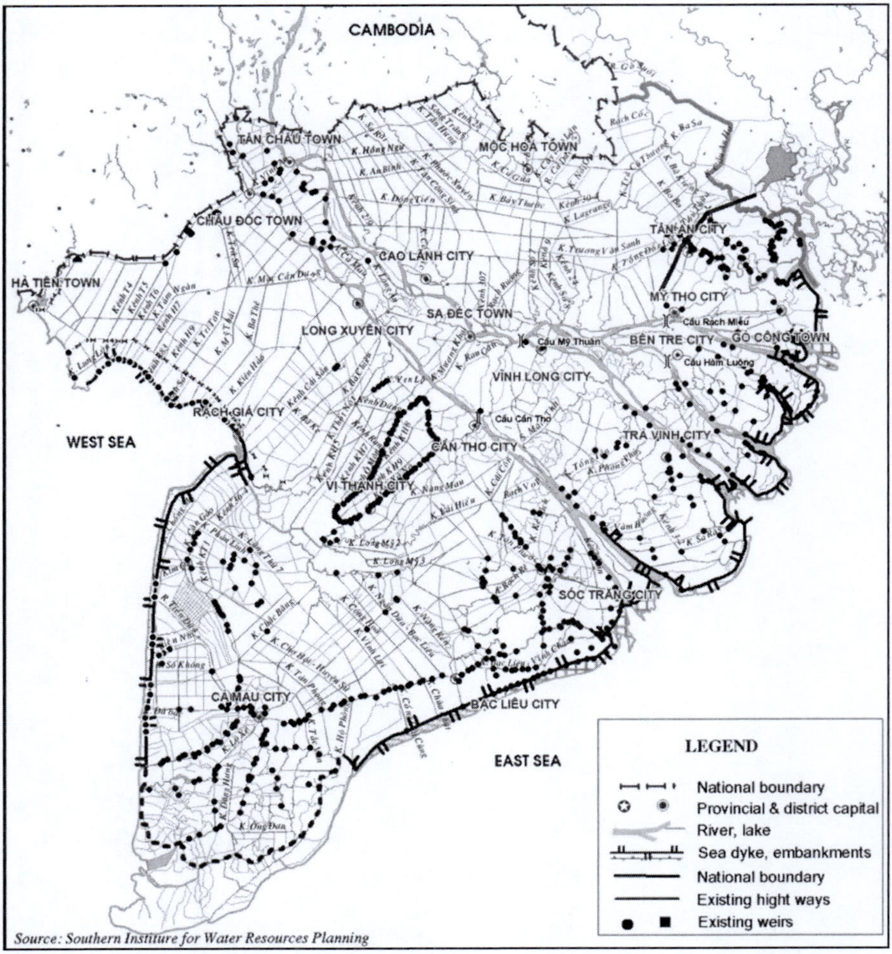

Source: Southern Institute For Water Resources Planning - SIWRP 2011b:16)

Delta's diversity of eco-hydrological zones

The Mekong Delta is characterised by a diverse environment of different ecologies. The natural, social, economic and political conditions are subjected to infrastructure planning and investments which drive the production choices (e.g. cropping patterns, cultivation practices) in individual zones. Based on physical, hydrological and agricultural features, the Delta is divided into 6 agro-ecological zones (Figure 3-4), excluding the hilly and mountainous areas (To

Van Truong 2002, cited by Pham Cong Huu 2011:45, Duong et al. 2005, cited in Nguyen Duy Can et al. 2007:75).

- Freshwater alluvial zone (the area between the Tien and Hau rivers)
- Plain of reeds (*Đồng Tháp Mười*)
- Long Xuyen (–*Ha Tien*) Quadrangle
- Trans-bassac depression zone
- Coastal zone (East Sea)
- Ca Mau peninsula[8]

For research and planning purposes, the Delta is also divided into zones based on the depth and extent of flooding. Pham Cong Huu (2011:45–46) offers a 3-zone classification:

- Deep flood area (most of the Long Xuyen Quadrangle, Plain of Reeds and the An Giang, Dong Thap and Long An provinces)
- Average flood area (most of Hau Giang, Vinh Long, Tien Giang and Can Tho; part of the Trans-bassac depression and freshwater alluvial zone)
- Shallow or no flood area (most of the coastal area and the Kien Giang, Ha Tien, Soc Trang, Bac Lieu, Ben Tre and Tra Vinh provinces).

These 3 areas receive inundation levels of approximately 2–4 m, 0.6–2 m and less than 0.6m respectively.

[8] See Table 1 in Southern Institute For Water Resources Planning - SIWRP (2011:15) for detailed information on water infrastructure in the Mekong Delta, divided into 4 zones: Plain of Reeds, Long Xuyen Quadrangle, Ca Mau peninsula and the area between the two rivers.

Figure 3-4: Eco-hydrological zones of the Mekong Delta

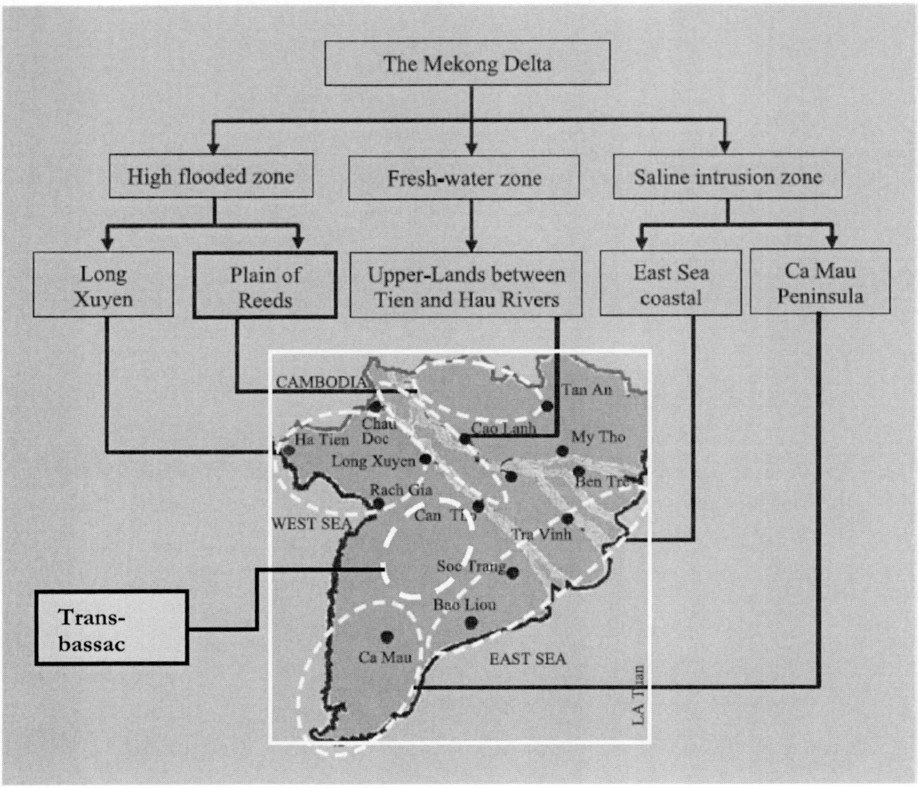

Le Anh Tuan et al. (2007:26), modified by the author according to Nguyen Duy Can et al. (2007:75)

Table 3-1: Six hydro-ecological zones of the Mekong Delta and their infrastructure and production features

Zone	Hydro-ecological regime	Infrastructure highlights	Production features
Freshwater alluvial zone	Fertile alluvial soil, year-round freshwater availability, modest flood regime	*(Includes part of the coastal zone between the Hau and Tien rivers)* Major irrigation systems: North Vam Nao (completed); South Vam Nao (mostly completed in 2011) with electric pumping stations (50–200 ha each) North Vinh An with small electric and oil pumps; North Mang Thit with embankments	Double and triple rice cropping, fruit orchards, upland crops, freshwater aquaculture
Plain of Reeds	Approximately 66% of the Delta's acid sulphate soils, extreme seasonal flooding, year-round freshwater availability	More than 50 large-scale pumping stations (built 1978–1990) irrigating 35,420 ha Small electric pump stations irrigating 100–200 ha each, constituting 20%–30% of the total irrigated area	Double and triple rice cropping, freshwater aquaculture (fish), melaleuca trees, pineapples, sugar cane
Long Xuyen-Ha Tien Quadrangle Zone	Acid sulphate soils, extreme seasonal flooding, modest seawater intrusion	Flood control: Cluster flood drainage system: sea dike RachGia-BaHon and drainage canals and flood gateways; flood control projects along the	Double and triple rice cropping, partly aquaculture (freshwater shrimps),

		Vinh Te canal; 23 new flood drainage canals from Rach Gia-Ha Tiencanal to the western sea Salinity sluices system in Hon Dat, Kien Luong and Rach Gia City (Kien Giang) 4 small reservoirs with a total capacity of 750,000 m^3 supply water for irrigation and 12,000 residents of hilly Tri Ton and Tinh Bien (An Giang)	upland crops and fruit trees Gravity irrigation or with pump
Trans-bassac depression zone	Modest levels of flooding and salinity intrusion, increased freshwater availability through the Quan Lo-Phung Hiep canal since 1996	*(See the Ca Mau peninsula)*	Dominated by double and triple rice cropping, fish farming and upland crops, sugar cane and pineapples
Coastal zone	Salinity intrusion, tidal inundation, freshwater scarcity, large area with acid sulphate soil, permanent freshwater ecology since 1998	Salinity prevention projects: dyke in Bao Dinh; in South Mang Thit: dyke projects on the Co Chien River (Cung Hau) and Bassac River (Dinh An)	Intensive shrimp farming (including rice-shrimp systems), fish and crab farming
Ca Mau peninsula zone	Salinity intrusion, tidal inundation from both East and West Sea tidal	(*Including the Trans-bassac depression zone and South bassac region of the*	Intensive shrimp farming (including rice–shrimp

	motion, scarce freshwater, peat swamps	*coastal zone*) Long coastal dyke Insufficient culverts preventing good drainage Pumping station: 3 power fixed-pumping stations with 7–8 pumps; 6 new electric pumping stations in Hau Giang, 2 pumps (450 m^3/hour) serving about 65 ha each, 15,000–16,000 small pumps in households Natural protection: Mangrove forests and *Melaleuca* trees[9] at the East Coast.	systems), partly rice–fish systems Mass transition from rice farming to aquaculture in 2000

Adapted from Benedikter 2014b:135, data from Duong et al. 2005, cited by Nguyen Duy Can et al. 2007:75-76, Southern Institute For Water Resources Planning - SIWRP 2011b.

* Note: The data for the hydraulic infrastructure highlights are taken from another zone categorisation system which divides the Mekong Delta into 4 zones: the Plain of Reeds, Long Xuyen Quadrangle, Ca Mau Peninsula and Trans-bassac region between the 2 rivers (see Figure 11 in Bucx et al. 2010:24). Therefore, the presented statistical data (e.g. number and length of canals, sluices and culverts) are only relative.

[9] *Melaleuca*, or tea tree, found in the coastal zone, can survive in an acidic environment and can reduce the acidity in water. It is used in projects to achieve sustainable reduction of acidity in water used for agriculture. Doing is important, as high acidity levels can threaten water quality (Duong Van Ni et al. 2001, cited in Fuhrmann 2008:13).

As illustrated in Table 3-1, the 6 hydro-ecological zones differ from each other in elevation, soil type, flood regime, water availability and hydraulic infrastructure interventions based on agricultural pattern. For instance, large–scale, closed irrigation systems with pumps allow the cultivation of triple rice during flooding in the freshwater alluvial zone and the Plain of Reeds, while dykes which protect against salinity are used in the coastal and Ca Mau peninsula zones, partly to create an environment for shrimp farming. Beyond regional variation, the areas present significant internal differences because of the micro-climate of localities. Even given similar natural conditions, different administrative units might choose separate investments, creating diverse physical conditions within a single hydrological area and changing cropping patterns. Thus, human interventions and changes in agricultural system not only depend on the natural conditions and existing physical and farming characteristics of area but also are significantly influenced by the state or ruler's policies. For instance, the Plain of Reeds and Long Xuyen Quadrangle would not exist as they are today — dominated by extended rice fields throughout the year — if the Vietnamese government had not made and enforced the audacious decision to '*tough the belly button of alum* [the core area of acid sulphate soil] *of the Delta*' by cutting though the area with canal projects to wash alum during the 1980s and 1990s[10]. In addition, state policies affect differently areas which have distinctive natural and infrastructure conditions. For instance, localities may apply to have 2 or 3 rice seasons per year or to engage in gravity or lifting irrigation.

The research sites are located in Can Tho within the Trans-bassac Depression. Similarly to other areas in this zone, the researched areas experience a modest level of flooding, and despite the industrialisation and urbanisation of the Delta's central city, a large part of the city is still devoted to double and triple rice production, which is the subject of this research. Despite the diversity in the zones' water landscape, the researched area is similar to other parts of the Delta through its inclusion within the state administrative system so that local practices are somewhat shaped by the interaction between the state and other stakeholders at the interface of the local level. The state has established and maintained a dense bureaucratic system of state agencies, ensuring its influence in all local affairs (see Chapter 4 for detailed illustration of the state structure in

---

[10] See 'Dấu ấn Võ Văn Kiệt trên vùng đất chín rồng' [The imprint of Vo Van Kiet on the land of Nine Dragon], Ngoc Quan 2012, http://vtv.vn/Thoi-su-trong-nuoc/Dau-an-Vo-Van-Kiet-tren-vung-dat-chin-rong/52234.vtv

the Mekong Delta). The diversity of the Delta and its impacts are also captured in the different natural and physical conditions of the researched sites.

In short, irrigation in the Mekong Delta is characterised by hydro-ecological diversity and different farming system according to the hydro-ecological regime and the available water infrastructure (designed and implemented by the state and local stakeholders) distinctive to and among zones. This section has discussed the inter-relations between the natural conditions, water control structures, farming system and livelihoods in specific localities and the diversity of the Mekong Delta in future planning and interventions. The present challenges of the Delta's hydraulic landscape derive not only from insufficient investment but also from the technocratic approach to transforming the Delta into a controlled water landscape, which has generated optimism, national pride and great uncertainty.

The technocratic approach and its disadvantages in the hydraulic water landscape of the Mekong Delta

The impressive changes in hydraulic infrastructure of the Mekong Delta are undeniable. However, in addition to benefits, the historical technological development which continues into the present has caused direct and indirect, negative impacts on the Delta's environment. These impacts include the general drawbacks of modern infrastructure and intensive agriculture and the results of a technocratic approach which prioritises technological solutions with little or no consideration of their social and management dimensions.

Within the inter-dependence of irrigation and agriculture, hydraulic infrastructure decides the choice of agricultural pattern and practices while meeting the requirements for rehabilitated or newly constructed irrigated agriculture, flooding and salinity prevention, canal system and other hydraulic works (e.g. dykes, embankments, culverts, sluices). The expansion of water infrastructure has enabled farmers to expand agricultural areas and intensify cropping season. During the 1990s, the most impressive change was the shift from mono-rice to 2 or 3 seasons of rice farming annually and the expansion of technologies innovation and transfers, including the intensive usage of chemicals in fertiliser and pesticides.

Modern intensive agriculture, while providing more food for the nation and more income for farmers, has also caused various issues in the Delta. Land preparation, fertiliser and pesticide use, animal production and aquaculture which causes non-point source pollution are deteriorating the Delta's water ecosystem.

> Due to increased pollution by these agrichemicals, the capacity of freshwater systems to provide essential ecosystem services such as to maintain agriculture and aquaculture productivity and sustain human health is therefore under great pressure. (Sebesvari et al. 2012:331)

In addition, the degradation of soil and water environments result from the intensification of rice and shrimp production in both alluvial and acid sulphate soils (ASS) (Vo Thi Guong and Nguyen My Hoa 2012:363). The Delta's biodiversity faces great threats from human encroachment, the conversion of original ecosystems to rice paddies, deteriorating water quality, the construction of dams on the Mekong River upstream from the Delta, overfishing and sea level rises caused by climate changes (Campbell 2012:311).

Particularly in the realm of hydraulic infrastructure, the benefits of technological development in the Delta's water system seem to have reached their limit. Despite efforts to control nature by modern water civilisation, the Delta's inhabitants and planners face historically persistent problems and various new challenges: (a) flooding of more than 1.9 million ha in the upper Delta; (b) salinity intrusion in more than 1.4 million ha in the coastal area; (c) acid sulphate soils and acid water movement in more than 1 million ha of lowland areas; and (d) shortages of fresh water for cultivation and domestic uses in more than 2.1 million ha.[11] (Southern Institute For Water Resources Planning - SIWRP 2011a:7). These challenges are human induced; that is, they occur only because of human interventions in nature and the increasing human demand for water for various activities. This section considers some examples of these issues.

First, flooding caused by natural fluctuations in the water level was not a problem until human settlement expanded throughout the Delta, even into flood-prone areas. The Delta's water fluctuation also is problematic for irrigated agriculture which requires inundation-free fields during the flooding period. Second, salinity intrusion has increased because of the falling water table caused by using groundwater to supplement shortages of surface water, especially during dry season. Third, acid sulphate soil emerged as a problem once cultivation and human settlement expanded into areas with such soil. Finally, the water shortage has resulted from several on-going issues, including seasonal fluctuation in rivers and rain serving as water sources (Pham Quang Tu et al. 2013:11), possible impacts from upstream interventions such as

---

[11] In the dry season, the average discharge of the Mekong River is less than 2,500 m3/s and even as low as 1,700 m3/s, with the groundwater table falling by 2–3 m in some places (Le Anh Tuan et al. 2007:25).

hydropower dams and water diversion projects, increasing demand for multi-purposed use of the Delta amid population growth and diversification of activities[12], and water pollution from domestic water use, industrial activities, aquaculture and agriculture.

Adding to the problems of modern civilisation, technologies themselves also bring negative, direct and indirect impacts on activities and inhabitants in the Delta. Notably, the closing of natural drainage or man-made canals with water gates for year-long cultivation is a leading cause of unnatural or artificial, higher, larger and longer inundations; the 2011 flood is representative of this problem (personal communication with a water expert, 2012). In a study on dyke planning in Can Tho city, Pham Cong Huu (2012:171) gives an overview of the impact of hydraulic infrastructure. In addition to the positive effects of flood control, dyke construction and intensification of rice production, hydraulic infrastructure in Can Tho has resulted in various drawbacks: water pollution, reduced soil fertility and erosion along rivers and canal (also mentioned by Vo Thi Guong and Nguyen My Hoa 2012:389).

Even when beneficial gains have compensated for the negative impacts of technologies, hydraulic works expansion in the Delta has been evaluated as insufficient and incomplete projects, and the structures operate at much less than designed capacity. Many of the 50 pump stations in the Plain of Reeds operate at 20% of the designed capacity. In addition, little or inadequate maintenance (dredging) leaves many canals with little capacity for water conveyance and local transportation (dredging) (Southern Institute For Water Resources Planning - SIWRP 2011a). Extensive infrastructure does not necessarily result in quality and technological progress, and project maintenance greatly affects efficiency. The problem of maintenance is analysed in the case of canal dredging in chapters 4 and 5. This analysis shows that inefficient management creates task slippage between state, communal and individual responsibilities and can cause macro-level problems.

Authorities and other stakeholders generally recognised these problems as their severity increases. Local residents took actions to stop the habits that they used to practice in everyday life of the Delta. For example, farmers stopped bathing in the river and limited domestic consumption of river water. However, the engineering–centred approach to irrigation and agriculture management

---

[12] The intensification of rice production is a main factor. Dry season rice cultivation has increased in the past 15 years, using 3.8 million ha — more than three times than in the other Lower Mekong Basin countries combined (Le Anh Tuan et al. 2007:19).

remains dominant due, in part, to the political ecology of the hydraulic mission. Since the return of large-scale irrigation system and hydraulic work in 19th-century colonialism (Molle et al. 2009:329), hydraulic missions with massive *corvée* labour and, later, machines and the technical and scientific knowledge of enthusiastic engineers had become a global trend. Subsequently, the new legitimacy of technical marvels (e.g. high dams, electricity) and the perceived unlimited power of science inspired by the mission to tame nature and make the deserts bloom. Irrigated farming became associated with civilisation and a *moral ideal of farming,* and '*hydraulic bureaucracies … had their secular priesthoods, acting in the name of the common good and in tandem with politicians and national leaders*' (Molle 2009, Wester 2009, cited in Molle et al. 2009:332). The trend continued into the 20th century, and despite the current shift in scientific ideology to environmentally and socially sensitive approaches, '*the spell of a hydraulic mission*' remains (Blake 2012:341). The dominant power and difficulty of breaking from the technocratic approach in the hydraulic water landscape arises from the existing system of benefits protected by 4 powerful actors — politicians, construction companies, landed elites and development banks (Molle et al. 2009:336) — likely associated in strategic groups in diverse, complex hydraulic societies (Evers and Benedikter 2009a, analysing the hydraulic societies of the Mekong Delta).

The Mekong Delta is part of the global trend of modern irrigation management. Although the state has considered reassessing or re-designing the hydraulic structures to improve the water landscape in all regions, including the Mekong Delta, no critical comments on the value and suitability which go beyond technical assessment of projects in order to question the traditional and current technocratic approach have been made. Biggs (2004:213) contends that, in addition to the modern global experience of population explosion and agricultural intensification, the unique political ecology of the water-reliant Delta is a significant factor in the continued preference to the mechanical approach. According to the author, the political ecology is illustrated by the late-1880s, steam-powered dredging machines which '*fitted the political and technical needs of the colonial state by replacing thousands of labourers needed for traditional canal or dike projects*' to the greatest benefits of the French; by the diesel-powered dredgers introduced by American companies in the 1960s for security and political reasons; and the present-day hydrocracies which '*seek to expand their budget and power and fulfil professional inclination towards infrastructure*' (ibid:214). This work shows that, while infrastructure continues to be a sign of state involvement at the localities, the irrigation structures are used as placeholders to define the boundary of acceptable local deviation and the level at which state authority must become dominant. The inter-relations and the political ecology of human intervention in the water landscape of the Mekong Delta are analysed and reviewed in the next section from a historical perspective, moving from the first Vietnamese migration to the South in the 18th century to contemporary state management under a market economy.

From an open to a closed water network: A history of the making of the hydraulic landscape in the Mekong Delta

> First, water; second, fertilisers; third, hard-working; fourth, seeds.
>
> -Farming proverb

It is unknown when this adage became popular in rural Vietnam. Today, although more factors are involved in the Mekong Delta, water, in the forms of irrigation and flooding, remains amongst the most important factors vital for agricultural production, particularly rice farming. Since the beginning of human settlement in the Delta, investing in water infrastructure to facilitate irrigation practices and avoid flood damages has been considered essential. As well as economic benefits, the construction of hydraulic infrastructure throughout the Delta's history has often had other purposes, including military and political ones.

As mentioned, human decisions about infrastructure investment are guided by natural conditions, selected farming systems and social and political agendas. Based on its diverse natural conditions and complex socio-economic and political history, the Mekong Delta has less in common with the traditional irrigated landscapes in Asia, such as ancient irrigated regime of Angkor upstream and hydraulic infrastructure around Long Xuyen (An Giang), and more in common with lands newly reclaimed lands by *modern* hydraulic bureaucracies, such as those described in Donald Worster's *Rivers of Empire* and Michael Adas's *The Burma Delta* (Biggs 2012:11). This section explains the various purposes behind massive infrastructure development, including nation-building campaigns and the *work-never-ends* paradigm of hydraulics works. The present water infrastructure is the result of a series of interventions built on top of each other or, in Biggs's (2012:10) words, the '*continuities and overlaps between precolonial, colonial, and postcolonial eras of nation building*'.

Before the South migration during the Nguyen Dynasty (pre-19th century)

The birth of the Delta is described as follows:

> In geologic terms, the valley formed over 2 million years ago in the late Tertiary period, and the first delta formed in what is now an arc of red-clay hills stretching from Hồ Chi Minh City to Tay Ninh and Phnom Penh. Even as late

> as 6000 BCE, well after the end of the last ice age, large areas of the modern delta area were still routinely inundated with seawater. Areas over sixty kilometres inland today contain traces of underlying coral reefs and sand dunes from the ancient coastline. (Nguyen Huu Chiem 1994, cited in Biggs 2012:14)

For the last 2,000 years, the young, alluvial Mekong Delta had rapidly expanded into the sea, driven by high loads of silt from the rivers (Nguyen Huu Chiem 1994, cited in Biggs 2012:14). From the very beginning of the Delta's life, human settlements were concentrated in elevated lands called *miệt vườn* (garden strips) along the river and ancient sand hills. Narrow settlements on the water's edge and forests, plains and mangroves made up the Delta (Biggs 2012:15, Son Nam 1997).

In this early period, the water infrastructure was associated with the pre-Khmer civilisation called Funan (ca. 300 BCE–600 CE) or Oc Eo Culture[13]. Occupying the territory of modern-day Cambodia and Southern Vietnam, the Funan people were the first to open the Mekong Delta by constructing an extensive network of canals for irrigation and transport (Weggel 2003, cited by Fuhrmann 2008:15). Thus, human investments in the water landscape of the Delta started in a very early period of settlement history and continued from the first Indianised civilisation of Southeast Asia to Chenla occupation (6th–8th centuries CE) and the Kingdom of Angkor. Likely, the water resources and monsoon climate which created opportunities for food production and water transportation made taming nature in the Delta a worthwhile project. In that sense, the position of the Mekong Delta in the Funan Kingdom was similar to that of the Tonle Sap lake in the Angkor kingdom (9th–15th centuries CE) (Kummu 2003:15).

In the following period, the Delta became the site of interactions between and interventions by various cultures separately or in combination, from the Khmer to the Chinese and then Vietnamese. According to Son Nam (1997:18), in the 13th century, the Khmer found news rice varieties and adapted the technique to *bring water to the field* (*dẫn thủy nhập điền*), which allowed 3 to 4 harvests a year. Since the late 1600s, the Delta has possessed a number of trading ports along

[13] Óc Eo, located in the modern-day An Giang province, was the region's major trading port. Funan is the name in Chinese dynastic histories applied to a maritime kingdom in the lower Mekong Valley of modern Cambodia and Vietnam. At the height of its power in the mid-3rd century, Funan is thought to have controlled some of the major ports on the Malays Peninsula and to have been influential in the development of maritime trade between India and China (Ooi 2004:529).

the main rivers, facilitating the exchange of goods and culture (Biggs 2012:56). In the 1700s, Chinese migrants, particularly the Minh loyalist (*Minh hương*) merchants, became the rulers of the semi-independent ports, which were expanded into Sài Gòn, Mỹ Tho and Hà Tiên under concession from the Vietnamese and Cambodian kingdoms (ibid:16–17). By the late 1700s, the backbone of the Delta's water infrastructure, with canals, forts, roads and markets constructed by the Khmer, Vietnamese and Chinese, has been established. By the early 1800s, the system of Khmer river towns had been greatly copied, expanded and modified by Nguyễn's military governors (ibid:59), showing the continuity in human interventions, one on top of the other.

The South migration (Nam Tiến) in the Nguyen Dynasty (late 1700s–1860s)

This section describes infrastructure interventions during the period of Vietnamese migration southward starting in the late 18th century. The first group of Vietnamese immigrants from Tonkin – the North of Vietnam – consisted of indigent farmers, poor craft workers, discharged or deserting soldiers, exiled prisoners even some rich individuals seeking new opportunities and fleeing heavy taxes, oppression, exploitation, epidemics and natural disasters under the feudal landlord class during the 175-year Trinh–Nguyen civil war (Bui Huy Dap and Nguyen Dien 1996:73, Son Nam 1997:8). These south-bound migrants attempted to conquer nature and the land, forming villages and establishing settlements on wild land (Son Nam 1997). The Vietnamese contributed to the reclamation of land in the Delta started by the Khmer and Chinese settlements. The South migration became massive when the Nguyen lord decided to move south to reclaim land in the 18th century. After years of civil war in the 1670s ended with Nguyen victory over the Trinh lord in the North and chaos in the Khmer Empire (Cao Miên), the Nguyen Empire decided to send troops to Sai Gon and the Mekong Delta (Son Nam 1997:29), leading to the 1802 proclamation of Nguyễn Emperor Gia Long as the ruler of the whole country, which was renamed Việt Nam (Weggel 1990, cited in Fuhrmann 2008:19).

The Vietnamese started settlements in the northern Delta along the 2 arms of the Mekong, with military posts (*đồn điền*) to secure central government control (Brocheux 1995:11). Until the mid-19th century, the Delta was described as a largely wild area, with newly established, floating rice fields in the sweet alluvial land between the Tien and Hau rivers and other separate zones. The settlers in this Delta came from a wide range of origins and established a new type of village different than the Northern cluster village protected by a village gate. Villages started in elevated areas and flat plain and gradually spread throughout the Delta, especially along canals whose banks were used as roads and living spaces (Bui Huy Dap and Nguyen Dien 1996:75). From the beginning, the Mekong Delta shaped human settlements with its natural threats (humans sought elevated areas to avoid flood and *poisonous water*) and opportunities (humans used canal networks to exploit water resources for everyday use, transportation and production). In turn, settlers have continuously changed the Delta's water landscape for various purposes. The dynamics of the Delta distinguish the South from the North and partially explain the dynamics of state and local management in the Delta.

In this period, hydraulic infrastructure was contrasted as interventions to connect the Delta internally and externally with the administration structure of the newly reclaimed territory. During the struggle with Khmer and Siam troops in the early 1700s, local projects of digging canals took place. Nguyen Cuu Van's troop constructed the Vũng Gù and Mỹ Tho canals in 1705 to connect the Vam Co Tay rivers though Tien Giang (Son Nam 1997:30). From the start, canals played an important role by providing major lines of transportation and communication to build the state in frontier land (Biggs 2010, Brocheux 1995, cited by Benedikter 2014b:82). After the establishment of Viet Nam as a nation in 1802, the kingdom invested in extensive administration and physical infrastructure throughout its territory especially in the 6 military provinces (*trấn*) of the Delta and Sai Gon because of the Delta's economic importance (ibid:18). In this period, the kingdom prioritised investments in 2 seasonally inundated floodplains: the Long Xuyen Quadrangle and Đồng Tháp (the Plain of Reeds), including 2 river ports, Chau Đốc and Long Xuyen, and 2 coastal ports, Hà Tiên and Rạch Giá (ibid:18). Even from this early period, the geopolitical importance of the Delta had resulted in extraordinary investment, above existing demand, to develop the Delta's resources.

In the 19th century, the Delta saw large-scale canal excavation and infrastructure rehabilitation, mostly for security and military purposes. The most significant projects — the Bảo Định, Thoại Hà and Vĩnh Tế canals — were constructed during the 18th to 19th centuries using recruited labour and soldiers to move Vietnamese settlements to the Cambodian border in order to, among other purposes, protect the Delta from future Siamese incursions (Biggs 2012:65-68). However, not long after completion, the canals were abandoned, also for security reason. After the civil tension in 1836, the canals became vulnerable to the invasions from sea and so needed to be closed. The security tension also shifted priorities in the next period from expanding the canal network to restoring the internal links from North to South. '*Every morsel of earth*' in the Mekong Delta was used to intensify farming efforts, while efforts to do so were enforced through taxation, education and land management (ibid:38). This first period of Vietnamese settlement in the southern Delta, with the empire's aim to acquire control over the new land, resulted in hydraulic interventions which served security purposes along strategic frontiers and enabled military transportation, although farming-oriented water projects also took place later.

French colonisation (1860s–1954)

Canal excavation for both agriculture and security purposes continued in colonial times but with a dramatic adoption of mechanisation. Replacing the use of soldiers and conscripted labourers to dig canals with simple tools, the French introduction of dredging machines, or dredges, opened a new era of hydraulic intervention in the Delta's water landscape. This period was highlighted by the transformation of water landscape under French colonisation, from French conquest in the 1860s through the struggles between French and Vietnamese military under Trung Trực's rule until 1868 to the war between French troops and the Viet Minh until 1954. The French colonial interventions employed modern scientific knowledge and technologies and a colonial administration equipped with engineers and planners to build a 'modern' hydraulics landscape (Biggs 2012:11). During the first colonial period, interventions served mostly transportation and military purposes and triggered land reclamation to colonise the Delta with structured institutions (Table 3-2, Biggs 2012:90, Brocheux 1995:17).

Table 3-2: Interventions in the canal system under French colonial rule (1860s–1954)

Year	Canal Projects	Funded and/or Constructed	Purposes
1860s	Canals in the dense, swampy forest of Bảy Thưa	Constructed by Trung Trực and his followers	Freshwater Communication with outside settlements
	Seven canals (~100m wide, 5–6m deep)	Funded by Paris, organised by the Department of Public Works (DPW)	Ship traffic through the eastern coastal zone
1876–1877	Chợ Gạo, or Duperre Canal — 11 km long	DPW	Connecting Sai Gon to My Tho Reducing travel time from Sai Gon to the Delta from 16 to 8 hours in protected water, instead of open sea Steamboat carrying passengers, mail and cargo across rivers to other Delta towns and upstream to Phnom Penh
~1891	Restoration of the Vinh Te Canal Ship canal parallel to the Chợ Gạo canal	DPW	Flushing basins and eliminating tidal dead points
1890–1930	Xà No Canal (1900–1904); a dozen other waterways, including the Quản Lộ–Phụng Hiệp	Steam-powered dredges	

	canal; canals in the Hau Giang region and Long Xuyen Quadrangle; Rach Gia-Ha Tien canal in 1930		
1885–1895	28 canals in Cai Be	Former district chief of Cai Be Trần Ba Lộc, owning over 2,000 ha., with hired labourers	Canals were later widened by the DPW
1896	Tổng Đốc Lộc canal — 10 m wide and 47 km across the Plain of Reeds Lagrange and Tháp Mười canals*	Trần Ba Lộc * DPW engineers following Lộc	Transportation
1903	Checkerboard grid of canals at 1 km intervals numbered as One Thousand canal (Kênh Một Ngan), Two Thousand canal (Kênh Hai Ngàn) and so forth; pumping stations along the Xà No canal	Remy Gressier of Gressier Estate	Supply water for 5,600 ha of land on both sides of the Xa No canal
1930	Rach Gia-Ha Tien canal 60-km long		Going through swampy, deforested terrain along the gulf coast, a 30-m-wide path carved through peat bogs, salt marshes and mangroves (ibid:87-88).

Made by author, data source: Biggs (2012)

The hydraulic interventions in the Delta during French colonisation involved a combination of central colonial policies and local efforts to open up the Delta to transportation and agricultural production. Under colonial policies, agricultural production is especially important for exports and, even more importantly, for controlling the Delta to the benefit of the French. However, despite following a French engineering agenda implement with new technologies, colonial interventions did not build a modern water landscape in an empty space. Rather, the French projects continued what the Vietnamese and Chinese had started. As reported by Renaud, the main 'painter' of French colonial hydraulics development, these were a '*work in progress*' to restore the waterways or to '*continue the work* [l'oeuvre] *of civilization started by the Vietnamese*' (Biggs 2012:5). Most significantly, this period witnessed the 1880 introduction of engineering technology to conquer nature: steam-powered dredges. Through the 1920s, the French excavated more than 1,500 km of navigable waterways and 2,500 km of secondary canals. This transformation of the waterscape sparked a new wave of Vietnamese southward migration (*nam tiến*), with an average 73,000 people annually settling along the canals from 1881–1921 and the moving outward to forests and swamps (ibid:71).

The French colonial ruler, especially after the 1879 French revolution, confirmed the commitment to canal maintenance[14], road creation, bridge construction, the study of railroads and the establishment of hospitals and schools as part of the general French policy to act as a benefactor to colonial territories (ibid:34). However, this ideology existed only on paper; French interventions continued to be politicised and dominated by the goal of controlling the colonial territory. The colonial economic system based on large farms and exports ultimately benefited only French nationals, mostly civil servants, and Vietnamese who were loyal collaborators with the French regime, excluding a huge group of poor tenant farmers (*tá điền*) (Brocheux 1995, cited by Fuhrmann 2008:24). Dredges became a technological tool for engineers and contractors, including the private ones who had been dominant since 1893, to gain power monopoly at the expense of local officials (Biggs 2012:45). (This situation was improved when the 1890–1930 canal building involved provincial planning (ibid). The problems of prioritising security purposes in interventions and using technologies as a means for certain groups to attain power and other

---

[14] The canal dredging projects in the Mekong The Delta were a test of the French reform *mission civilisatrice*'s ability to achieve canal building improvement, colonial modernisation and labour reforms (the abolishment of *corvee* – forced labour) using new science and technologies (Biggs 2012:35). However, the reform of paid labour failed because of changes in local dredging project. Much of the money went to labour contractors and other middlemen, and the government could not afford to pay tens of thousands of workers, even at low rates.

benefits are not specific to the French. These strategies were carried out during the pre-colonial empire and continued on in the following period of American colonisation and in post-colonial period. However, the degree of prioritisation and allowance for other goals (e.g. social welfare, agricultural production for the benefits of others) varied by political regimes and periods.

Despite of the power and benefits gained through colonial control and technology monopoly, engineers under the French colonial government faced difficulties stemming from both nature and human opponents in the Delta. The tidal and swampy environment challenged the French rulers with silted canals, acidic and salty soil and destructive floods and hindered canal interventions with new flooding and inundation zones[15], alum contamination[16] and lower water table[17] (Biggs 2012:72). While effective for French engineers, the new modern technologies were in all. For instance, the flushing basin technique which eliminating the dead points in the Cho Gao and Saintard canals served as a solution for engineers but prompted resistance from farmers, which forced French authorities to restore water circulation vital to farmers' survival (ibid:41). In another case, locals became angry with the Xa No canal project when the 3-m clay wall blocked natural waterways, creating inundation on one side and drought on the other (ibid:76).

Despite the drawbacks and inequality created by the colonial regime, the canal projects and other hydraulic interventions in this period laid a foundation for a controlled water landscape and shaped the Delta's subsequent human settlements. The pumps and water gates, which were the first of their kind in the region, impressed local residents, and the checkerboard canal grid built in 1903 became the foundation for the establishment of hamlets named according to the canal: One Thousand canal, One Thousand hamlet and so on. This period saw the advent of 'high modernism' (Scott 1998) as the French used the legitimacy of science and technology to evoke faith in their ability of their ideas to achieve their interests. The ideology of modernism has persisted in the Delta and aligned with global modernisation.

---

[15] For example, the Quản Lộ–Phụng Hiệp canal (1918) created new flooding issues in the Ca Mau peninsula (ibid:95).

[16] The new canal network reduced tidal fluctuations, preventing farmers from draining acidic water from their fields.

[17] The lower water table in the Long Xuyen Quadrangle caused by new canal excavation increased the exposure of the soil to sun, produced more acidic sulphate through oxidation and allowed salt water to move further inland (ibid:96).

During the balkanisation period of 1945–1954 (Biggs 2012:127), the Democratic Republic of Vietnam was established in the North and started nation building. The South and Mekong Delta, in particular, witnessed the struggle for power and control by various groups: the French, Viet Minh, Hoa Hao and Cao Dai troops[18]. These struggles led to the First Indochinese War. Interventions in waterways during this time were few, with isolated projects to serve military purposes. In 1946 and 1947, '*the water landscape gradually split apart (socially and hydrologically) into separate water enclaves: the one of the Viet Minh and the one of the French*'. Through 1949 on the Viet Minh side, the People's Army canals (Kinh Quan Dan) altered the waterscape to allow for quicker travel from U Minh eastward to coastal bases on the South China Sea (Biggs 2012:138). Canal dredging was accelerated from 1948 to 1954 under the guidance of Lê Duẩn, (head of the Party's Southern Regional Committee (*Xứ Ủy Nam Bộ*) and other Northern party leaders. One outstanding project was the 11-km People's Army Canal constructed in the dry season of 1949 to transport people and material in and out of the Delta near Bạc Lieu along the Sea (Biggs 2012:145). In the territory under French control, the water network was rehabilitated in early 1947 to protect convoys transporting equipment and rice along the Delta (Biggs 2012:138). Although the struggle destroyed roads and bridges, the Viet Minh did play an important role in regenerating the wetlands of the Delta (ibid: 139). The stagnation in infrastructure development due to warfare changed slightly with the official presence of US troops in the Delta around 1954, however, the Delta entered the destructive American War.

Thus, under French colonisation, the water landscape was strongly influenced by the ideological struggle to develop a colonial economy to benefit the colonial power and colony or to maintain security and control over the colonial territory. The later purpose, which resulted in politicised hydraulic interventions, dominated in this period. Despite the French interventions, the natural conditions, social and political conditions and conflicts over inequality made achieving colonial goals difficult for more than 100 years. Therefore, '*the French colonial administration never managed to gain full hydraulic control of the Mekong Delta*' (Evers and Benedikter 2009:13) but did succeed in establishing the ideas of a hydrological bureaucracy.

[18] From 1947 to 1950, the three Vietnamese groups expanded their control over portions of the Delta, establishing their centres of power known as Cao Đài Tây Ninh, Hòa Hảo Lang Linh and Việt Minh Can Gao (Biggs 2012:138).

The American war (1954–1975)

After the French regime, South Vietnam and the Mekong Delta, in particular, went through a Green Revolution with the introduction of new technologies. Water pumps became the most popular tool in rural Mekong Delta, as they remain so today. However, as under French colonialism, the hydraulic and agriculture interventions in the Delta during the American war were politicised. The consequent struggles over social inequality led to the termination of interventions in the Delta, especially amid the war escalation in the 1960s.

Immediately after taking control of the Mekong Delta in 1954, American advisers drafted large-scale plans to close off the entire Delta to realise its full agricultural potential (Benedikter 2014b:171). The Washington-Sai Gon coalition and Ngo Dinh Diem's government aimed to spread the Green Revolution and improve the social conditions through the Mekong Delta Development Programme (ibid: 86). The program modernised and mechanised agriculture with high-yield rice varieties and agro-chemicals, implemented agrovilles with dykes designed to protect settlement clusters from flooding, and introduced new irrigation systems, agricultural machines and pumps and dredges for silted canals and new waterways.

In general during this period, the American rulers' investments were strongly influenced by the political agenda to secure control in face of the obvious risk from the opposing military force – the Viet Minh. This overriding need is especially clear in the policy of agrovilles (*khu trù mật*) and strategic hamlets (*ấp chiến lược*). In 1957, the Land Settlement General Commission (*Tổng-Ủy Phủ Dinh Điền*) was established to take responsibility for reclamation and settlement projects, including 4 special projects at Cái Sắn, U Minh, Đồng Tháp, and the central highlands (Biggs 2012:169). Next, the Agricultural Mechanisation Campaign (*Quốc-Gia Nông-Cụ Cơ-Giới-Cuộc*) brought in hundreds of tractors and tillers and thousands of newly arrived diesel engines used as outboard motors and water pumps (ibid:170). The period also witnessed the involvement of French, Japanese, Israeli, Taiwanese, Dutch and other nationals who joined Americans and Vietnamese in Sai Gon as consultants working on individual contracts. The US, United Nations and foreign embassies[19] provided funds

[19] The interventions in Vietnam during this period was also influenced by the international discourse on modernisation, including the promotion of public–private partnerships for the design and construction of dams and other hydro projects, the collapse of the mainland Chinese Nationalists in 1949 and the creation of international

(ibid:72). However, despite the funds, equipment and technologies introduced, Diem and his ministers, not the Americans, were the main architects of canal projects, such as Cai Sắn, and later programmes, such as Dinh Điền settlements, agrovilles (*khu trù mật*), and strategic hamlets (*ấp chiến lược*). While technology and resources might have come from outside the country, the hydrological bureaucracy remained a firmly Vietnamese or local effort.

As same during the preceding period, American colonial actions often hid inequality, giving benefits to certain group at the neglect of others. For instance, the agrovilles were designed specifically for occupation by loyal Catholic refugees escaping from the communist North. In addition, as an unavoidable condition of war, the ambitious engineering plan of the American and their followers faced guerrilla attacks and escalated military conflict. These engineerings then were limited to small, isolated projects, and more often than not, the projects did not proceed smoothly to completion, as had happened in the Nguyen era and under French colonisation (Biggs 2012).

On the other side of the frontline, the Viet Minh also established their own military canals for covert transportation. For example, the Resistance canal (*Kinh Kháng Chiến*) in Mỹ An, close to Cambodian border, was a 'secret canal' running 20 km and connected by 2 major creeks in western Đồng Tháp to a river that formed part of the Cambodian border (ibid:178). After Diệm's era, Nguyen Van Thieu's government conducted a land survey in 1968 to expand its control in Dong Thap; however, the plans were not carried out (ibid:216). Interventions in water infrastructure in the later period were stopped by the acceleration of the war, which ended with the complete withdrawal of the American troops in 1975 and by the reunification of North and South Vietnam under Communism.

In short, interventions during this period were highly politicised, and the escalation of war reduced the effectiveness of hydraulic projects. However, the American War period was marked by specialised agricultural innovations. Alongside investments in military technologies and construction, Americans spurred a revolution in everyday technologies, such as small engines[20] attached

---

agencies to oversee regional and river basin development schemes (Biggs 2012:172). Consequently, the Mekong Committee was established in 1957.

[20] The motors are named *máy kô-le* for the American company and town (Kohler, Wisconsin) in which they were made. Diesel engines attached to sampans could quickly transport travellers beyond government checkpoints into National Liberation Front base areas. When not used for transportation, the engines served as water pumps

to boats and pumps for irrigation in the Delta. Biggs asserts that '*more than the violence or spectacular displays of military hardware, the proliferation of boat engines, radios, and motor pumps played a central role in the delta's environmental history*', and '*such machines were a matter of life and death for many in the countryside*' (ibid: 200), achieving success that Nelson called the '*demand theory of invention*' (Sansom 1969:177). An estimated more than 1 million pumps were in operation by 1974 (ibid:209). The pump and the dredge survived as technologies in the Mekong Delta much longer than the colonial regime because of their powerful benefits and assistance in the human conquest of nature. The number of pumps in the Delta has increased over time, rising from approximately 150,000 in 1975 to more than 350,000 in 1999 (GSO, cited in Benedikter 2014b:68). In the 2011–2012 interviews and observations conducted in the Mekong Delta, nearly all the farming households owned at least 1 pump (this research was conducted in Can Tho, Bac Lieu and An Giang). Access to technology, thus, facilitated independence in farming, contributing to the more capitalistic and individualistic lifestyle in rural Mekong Delta.

As in the preceding period, interventions under American colonialism were constrained by nature and locally established social arrangements. The 1950s canal projects in Dong Thap faced the same obstacles as the French had: fragile, flood-prone wetlands; salty water; parched, peat soil; poor knowledge of local conditions; and resistance from local Viet Minh supporters (Biggs 2012:174). Consequently, the land became a real 'quagmire' for nation builders (ibid:179). For example, the salt barriers in Soc Trang, built in 1957 with financial support from the World Bank to protect 55,000 ha of farmland from salt intrusion, created stagnant freshwater behind the dam, which was mitigated by a local pumping initiative. Ultimately, the nature of the Delta guided and shaped colonial investments in canal networks and agricultural technologies. In particularly, French colonial canal designs[21] and earlier Vietnamese frontier settlement initiatives, such as the Southward march (*nam tiến*) and agricultural garrisons (*đồn điền*), became the main contributors to the draft of agrovilles and guided the design of American interventions (Biggs 2012:155). Thus, the American colonial period saw continuity in interventions in the Delta landscape,

---

(creatively adapted by Phạm Văn Thanh, a small-engine merchant in Mỹ Tho and later the town's richest resident who sold an average of 600 new pumps and motors a month through mid-1967) to individually manage water levels within paddy dikes. The booming of engine sales in the Delta was a rare major American success in the Delta (Biggs 2012:207–208).

[21] Nation building after 1954 was based on reports from technical surveys and field data from the French DPW produced starting in the 1880s (Biggs 2012:163).

with new interventions built on infrastructure and arrangements established in preceding periods.

Nation-building after reunification in 1975

The post-colonial period has been characterised by one ruling system and a specific political regime, with periodic uncertainties and outside threats. Nation building in the Mekong Delta since 1975 falls into 2 periods with different primary purposes of state interventions. In 1975 and 1976, the Communist state invested in hydraulic infrastructure to serve security purposes (to control the newly reunited country), and the projects were strongly aligned with collectivised agriculture. The second period of nation-building after Doi Moi was marked the construction of modern hydraulic infrastructure intended to tame nature and serve in the expansion, intensification and modernisation of the farming system. As well, technological investments driven by high modernism ideology have been aimed at retaining the privileges and benefits of irrigation management to related agencies or hydrocracies.

Immediately after reunification in 1975, the Vietnamese government undertook a nation-building campaign in the South, including the Mekong Delta, intended to (1) restore the agriculture, land and economy damaged by the war; (2) integrate the South into the Communism regime that had been established and developed in the North since 1945; and (3), most politically important, control the area which had been under different rule for 25 years. The nation-building started with the restoration and excavation of waterways to facilitate agricultural collectivisation. However, the policy of collectivising the input and output of agriculture production led to a stagnant economy, mainly because of the low incentive to invest and contribute to collective farming. The extreme crisis during the late 1970s and early 1980s prompted reforms in the late 1980s (Renovation, or Doi Moi in Vietnamese). Throughout changes in political regime and policies, infrastructure building, especially for irrigation, has remained a central goal of intervention in the Southern Delta, as illustrated by large-scale irrigation systems, canal maintenance projects and dyke systems to mitigate floods and intensify agricultural production (Southern Institute For Water Resources Planning - SIWRP 2011b, Benedikter 2014b, Biggs 2012, Kono 2001, Käkönen 2008).

The obstacles of the complex Delta have persisted and continued to challenge the efforts to control it in the post-colonial period. As stated by Benedikter (2014b:232), '*many of today's problems in the delta are newer versions of much older problems and tensions in the region*'. At the same time, the diversification, expansion and intensification of agriculture amid globalisation have pushed the Vietnamese state to further invest and modernise the water infrastructure of the

Delta. In the late 20th and early 21st century, the nation-building of hydraulics infrastructure has been characterised by several features.

First, the government's top-down policies [22] of modernising irrigation infrastructure focus on hardware interventions — constructing and maintaining hydraulics works (i.e. canals, dykes, sluices, culverts). Influenced by global ideologies and beliefs in recent decades, Vietnam has followed the past, current and persistent worldwide trends in its efforts to control the Mekong Delta's water landscape by applying the technocratic approach.

Second, post-Doi Moi decentralisation of state management has paved the way for dividing between levels of the state tasks in the construction, operation and maintenance of hydraulics works. Despite the decentralisation of state management, scholars working on the Vietnamese irrigation sector argue that the central state neglects the role of other stakeholders, including provincial and lower-level government officials, private enterprises, research institutes and individuals (Benedikter 2014b:235, Pham Cong Huu 2011). This neglect increases the risk of continued unawareness of potential environmental and social impacts of water infrastructure projects, which might lead the state to undermine the importance of potential impacts and their relation to political stability. The consequent effects on state involvement in day-to-day management and local views of the state are highlighted in this analysis. As well, the combination of decentralisation and local, centralised state management in hydraulic operation and maintenance (further analysed in Chapter 4) has *'created space and mobility for the state engineers and water bureaucrats in Hanoi to seize the southern waterscape and expand its power'* since the late 1970s. Consequently, the Northern hierarchical apparatus of water management agencies (the hydraulic bureaucracies, or hydrocracies) was copied in the Southern Delta, with human resources transferred from the North to occupy important positions (ibid:45-47). However, since Doi Moi, the equitisation of state-owned enterprises and the shift to the socialism-oriented market economy in Vietnam, the growing number of private enterprises has diluted the monopoly power of state-owned construction companies in hydraulic projects (Benedikter 2014b:228).

Third, following the nature of local deviation throughout Delta history, local resistance and arrangements can affect the implementation of central plans. In particular, the Delta's natural, physical and social conditions continue to influence state interventions to various extents depending on the activities and

[22] These top-down policies are supported by the Leninist doctrine of statehood and the rationale of central planning (Benedikter 2014b:32).

locations. As early as the 1990s, the Mekong Delta entered a new era of engineering shaped by regulatory devices (Miller, cited by Benedikter 2013:88). Benedikter (ibid:222) asserts the post-colonial era has seen 'the shift from '*opening up'* the Delta environment by canal dredging to *'closing it off'* by the development of a large-scale regulatory infrastructure', or in another words, the transformation of a *river-water civilization* to a *modern hydraulic society* (Fuhrmann 2008) in which almost every inch of the Delta's surface shows signs of human interventions (Biggs 2012:6). However, the Northern model of hydraulic control proved to have many shortcomings when applied to the Mekong Delta and its specific natural and social conditions. As a result, the contemporary Delta stands as a regulated water landscape with lower level of hydraulic control than in the Red River Delta, inter-connected canals with minimal regulatory input and only a few large-scale, closed hydraulic systems [23] (Benedikter 2014b:81, Vo Khac Tri 2012:74).

Fourth, while guided by the national state plans, the present-day water landscape of the Mekong Delta has since the 1990s been significantly influenced by technical and funding supports from international donors, including development banks and the Dutch, German and Australian governments. These projects tended to favour technocratic solutions. However, since the 1990s[24], hydraulic experts worldwide have condition how to better plan the hydraulic works by combining technocratic and management approaches to water management (Renault et al. 2007). Making participation by actors from the central to local levels has become more viable and effective (Pham Cong Huu 2011) and assessing the impact of water projects have gained attention only after severe, natural or man-made disasters (Benedikter 2014b:229).

Above all, the hydraulic sector is strictly controlled by the '*synergetic relationships*' among water bureaucrats, state and local politicians, water enterprises and development banks (Molle et al. 2009:336). Thus, '*the ways the flows of water are created or modified by water infrastructure are intertwined with flows of power and influence, often manifested in the form of political or financial benefits, whether private or collective*' (ibid). In the Mekong Delta, global trends have been translated into multi/interdisciplinary projects. Most notably, in the Mekong Delta Plan launched in 2010, the Vietnamese government and Dutch agencies are

---

[23] See map 3.1 in Benedikter (2014b:81) for the closed hydraulics plans in the Mekong Delta, most of which have not been completed.

[24] The dominant ideologies have been irrigation management transfer, water user groups and, finally, integrated water resources management.

collaborating '*to create a strategic long-term vision (2100) for the Mekong Delta, to set out a long-term Delta Program with a range of short-term measures (2015-2025) and to strengthen water governance for the delta*'[25].

In summary, post-colonial nation-building reflects trends in human intervention in the water landscape of the Mekong Delta persistent before the pre-colonial time. The trends include restoring old projects, expanding canal network and applying *modern* technologies in water infrastructure (e.g. dykes, sluice, culverts, large-scale irrigation systems) to serve either political aims (territorial control) or economic purposes (agricultural production). Despite differences in forms, the post-colonial era has repeated mistakes from earlier periods, specifically using water project architects unfamiliar with the Delta's natural and social conditions, although the extent of trend has abated recently. The projects undertaken immediately after reunification clearly illustrate this tendency. In 1975, the Vietnamese government implemented a hybrid of the Green and Red revolutions seeking to make the Delta's waterscape the '*agrarian and irrigation frontlines*' for the Party, military and people[26] (Benedikter 2014b:42). However, Northern engineers' lack of knowledge about the Delta and insistence on achieving certain targets resulted in the failure of the ambitious, capital-intensive venture to build a network of electric pumping stations and dyke polders during the 1970s and 1980s (Benedikter 2014b:50, To Van Truong 2011). Consequently, the Vietnam government has applied the *trial–and–error* principle to water infrastructure projects in order to responding to the Delta's hydrological and social requirements (Benedikter 2014b:50). Despite the changes in attempts to conquer the Delta's water landscapes — from soldiers, conscripted colonial subjects and comrades digging to large machines run by state-owned and private companies, human interventions in the Delta have remained '*work in progress*' (Renaud's words, Biggs 2012:5) or '*work without end*' (Bigg et al., cited by Benedikter 2014b:94). Water infrastructure in the Delta represents '*an ongoing and seemingly endless engineering campaign*' (ibid:94).

---

25 http://www.deltares.nl/en/news/news-item/item/13775/dutch-expertise-mekong-deltaplan-vietnam, accessed 07.03.2015

26 The canal projects were organised as campaigns, and a military victory-like celebration was organised for each completed canal project. Important days and their names were used to inaugurate and name the new canals (Benedikter 2014b:42).

Agricultural and rural transformation in the Mekong Delta

Closely associated with and dependent on irrigation infrastructure, Delta agriculture has been significantly affected by interventions in the water landscape. Simultaneously, a rural society with a rich, diverse culture closely attached to water and farming has been established even amid continuous transformation.

Agricultural transformation

Through various periods of nation building and intervention, Delta agriculture has undergone dynamic transformations, highlighted by the shift from wild rice collection and floating rice farming to irrigated agriculture. The transformation was initiated by the first intensive canal excavation in the Nguyen era, continued with the first mechanisation of agriculture with French technologies, which was extended into the colonial American period dominated by the import of individual pump Kô-le and agro-chemicals. Traditional practices of floating rice farming, returned during the military escalation in the late 1960s, gave way to the modern hydraulic agriculture of the present day. The contemporary farming system has resulted from interventions in agriculture supported by water infrastructure campaigns since the 1970s. Highlights of the transformation include the increase of cropping seasons and the application of new rice varieties with higher yield and short growing time. Details of the historical agricultural transformation are presented in Table 3-3.

Table 3-3: Evolution of agrarian systems in the Mekong Delta

Period	Farming Pattern	Technologies
Rice farming in the Oc Eo period and under the Nguyen dynasty (roughly before 1857)	Collection of wild rice for food and cultivation of floating rice[27]; Rice cultivation along the canal system established in the Nguyen era.	Natural irrigation, drainage and alum washing, depending on tidal system; 1800s - Vietnamese migrants borrowed Khmer tools and work animals to deal with marshy soil and cut grass.
Traditional rice farming system (roughly 1857–1966)	Tenant farming; Floating rice in deep-water flood areas with single and double transplanting of rice; High-yield varieties (IR5 and IR8[28]) introduced in 1966.	1860s, double transplanted[29]; Late 1860s, changed to direct seeding thanks to the plough; Colonial French - machinery, motorised water pumps, and

[27] Floating rice is closely related to a perennial wild species of rice common in marshes. Floating rice does not actually float, but its stem grows rapidly up to 4 m, keeping the rice above the water during flooding. The flood-tolerant, long-stem rice served the interests of Khmer farmers, while Vietnamese farmers preferred the higher-yield, short-stem varieties (Biggs 2012:63).

[28] IR8, released by the International Rice Research Institute in late 1966, became the most widely grown variety in a couple of years, making the Green Revolution in Asian rice. The variety was named the miracle rice, Rice of the Farming God (*Lua Than Nong*) and Lua Honda (Honda rice) because one good crop could make enough profit to buy a new motorbike (Hargrove 2006:40-41) http://www.adron.sr/files/RiceToday-2006-10-vol05-nr04.pdf, accessed 19.11.2014.

[29] This process started in May when rain came to the tidal-affected floodplain. Seeds were sown in the nursery beds in the lowest parts of the rice fields (~5% of the main fields). Weeding occurred in June, and when the rain fell regularly and the soil became soft, the seedlings were transplanted to the second nursery bed. In August, all the paddy fields were covered with 30–40 cm water (up to 70–80 cm) from continuous, heavy rainfall.

		tractor-pulled ploughs but only at large-scale plantations[30]; 1960s, Americans introduced pumps and agro-chemicals. Third season started.
Transition from traditional rice farming to high-yield rice farming (roughly 1966–1975)	High-yield rice (IR5 and IR8) - Double rice cropping – moving to commercial agriculture; From 1968, due to war escalation, return to wild rice, phantom rice (*lúa ma Oryza rufipogon*) and wild fauna for survival.	Machines were used for land preparation and irrigation.
Rice-based farming system development (roughly 1976–1988)	Double rice cropping: winter–spring rice, and rainy season rice or summer–autumn rice; Collective farming and, later, contract farming until Doi Moi announced in 1986; Varieties[31]: Crossed short-stem strains introduced from northern Vietnam and China with long-stem strains favoured by Khmer farmers; short-stem rice (*Oryza sativa*), including non-glutinous (*lúa tẻ*) and glutinous (*lúa nếp*) types; floating rice (*Oryza*	Harvest tools were, for short-stem rice, a short-handled sickle, and for floating rice, a small blade attached to a longer and curved wooden handle[32]; Khmer oxen and water buffalo which better tolerate the boggy soils were used for ploughing and transporting;

[30] These include the French plantation at Co Đo, the lands of Huỳnh Kỳ, the Gressier Estate o then Xa No Canal and other large French plantations in Bạc Lieu.

[31] In 1990, more than 500 varieties included 16 varieties of floating rice grown in the Delta differentiated by adaptability, length of growing season, taste, colour and size. These were named for local language and customs and reflected poetic, religious or practical themes (e.g., White Phoenix, 'Bird Claw, Sweet Girl)

[32] Handles were made from mostly the resinous tree *cây mù u* (*Callophyllum inophyllum*).

	rufipogon); 1979 – Innovation of the brown plant hopper-tolerant rice varieties.	From 1979, collective farms were established.
Intensification of the rice-based farming system (roughly 1988-1999)	Intensification and diversification (including upland crops: bean, corn, sesame, shrimp, fish culture and livestock); Increase triple cropping – commercial rice production; IRRI's HYVs rice (< 100 days of growth and harvest about 8–10 tonnes/ha/year – double or triple cropping)[33].	Since Doi Moi, introduced contract system and decollectivisation; New rice-growing methods; Integrated pest management was implemented starting in 1992.
Sustainable (ecological) farming system (since the early 2000s)	Term 'sustainable agriculture' used in government policy since the late 1990s; New government policy on 'change in the economic structure of agricultural systems' released (2000); incorporation of agribusiness management for rice production (2001).	Individual farming, pumping, ploughing, harvesting, machines and sales.

Made by author with data from Nguyen Van Sanh et al. (1998), Nguyen Duy Can et al. (2007:78), Kono (2001), Biggs (2010), Benedikter (2014).

Recently, double and triple rice farming have become dominant in the practice of rice cultivation in the Delta. In areas where triple rice farming is practiced,

[33] With high-yield varieties, Vietnam went from being a net rice importer in the 1980s to become the world's second- or third-largest rice exporter in the 1990s. The yield continued to rise, along with the intensification of rice cultivation (2–3 rice seasons/year), increasing the Delta's rice production from 4.5 million tonnes in 1976 to nearly 21 million tonnes in 2008 (SIWRP 2011:7; see Benedikter (2014b) for data on the expansion of rice cultivation areas and production from 1976 to 2009).

the timing of seasonal planting is planned carefully (Table 3-4). Thus, the contemporary agricultural system has made human settlements more dependent on the hydraulic infrastructure to sustain their income through agriculture production and the export sector.

Table 3-4: Standard schedule for 3–rice cropping

Rice season	Schedule	Irrigation Practices
Winter–spring (*Dong Xuan* – 1st season)	Nov/Dec–Feb/March	Field drainage (gravity or by pump) is done before sowing.
Spring–summer (*Xuan He–sạ chay*)	Feb/March–June/July	During dry weather in February or March, rice straw is spread over the fields and burned. The fields are then immediately irrigated to the depth of 5–20 cm, and seeds are broadcast. A day or 2 later, the water is drained, and the fields are left dry for several days. Then, chemical fertilizer and herbicide are applied, and the fields are irrigated again. A water level of 5–10 cm is maintained until the harvest.
Third season (He Thu or summer–autumn season)	June/July–Aug/Sept[34]	Immediately after the harvest of Xuan He season in June or July, it is crucial that farmers decide whether to attempt a third rice crop. If so, they immediately start the broadcasting in order to avoid flood damage at the harvest time.

Source: Kono (2001:82)

Considering the agricultural transformation, Kono (2001:83) concludes that the '*rapid interaction between infrastructure improvement and technology innovation accelerated the evolution of paddy based farming in the Mekong delta and intensified the cropping system*

[34] Considering time limits, farmers select IR50404 for its short growing time of only 85–90 days.

within a short period'. Investments of hydraulic structures in the water landscape, regardless of their primary purpose, provide foundation for agricultural transformation. The pre-colonial and colonial periods clearly demonstrate this claim through the expansion of the canal network before and during the Nguyen era and the introduction of dredges by the French and of pumps, agricultural practices, and chemicals by Americans. A similar relationship between infrastructure change and agriculture practice has occurred in the post-colonial era. For example, dyke construction in 5 zones of Can Tho city facilitated the change from 2 to 2–3 rice crops and from extensive to intensive fish and shrimp culture (Pham Cong Huu 2011:84). In addition, the agriculture transformation and targeted plans have required the expansion and the rehabilitation of the hydraulic infrastructure. As Kono (2001:77) asserts, the expansion of double rice cropping in the 1970s and 1980s would not have been possible without upgrading the canal system to open access to water resources in order cultivate rice in more areas.

Today, the Mekong Delta continues to undergo diversification and intensification of heavily mechanised agricultural production. While moving towards industrialised, market-oriented agriculture, the agricultural planning for the Mekong Delta through 2020 aims to keep 3.7 million ha under rice cultivation, giving priority to commercial production at large-scale farms supplying rice for processing centres in Long Xuyen Quadrangle, the Plain of Reeds and Ca Mau Peninsula (MARD 2009:28). To facilitate intensive rice production, mechanisation has been considered a solution for large-scale commercial farming. Regulated through legal documents (Decisions 63/2010/QĐ-TTg and 65/2011/QĐ-TTg), the Vietnamese government encourages individuals and organisations to invest in machines (e.g. high-capacity pumps, ploughs, harvesters, dryers) throughout all stages of rice farming, etc. with loan support from the national banks. According to the Department of Processing and Trading (2012), the Mekong Delta has 12,455 harvesting machines, including 8,919 combined harvesting machines and 3,536 line spreading reapers, and agricultural production in the Delta is 80% mechanised in soil preparation, 30% in harvesting, 42% in active rice drying and 95% in both threshing and milling. Most provinces have mechanised rice production on 50-60% of rice cultivated land.

Investment in hydraulic infrastructure is equally important not only for irrigation but also for transportation, flood mitigation and water supply. The prime minister Nguyen Tan Dung issued decision 1397/QĐ-TTg approving the irrigation planning for the Mekong Delta for 2012–2020 and through 2050 based on expected climatic changes and sea level rise. In this recent plan, the government continues to prioritise investments in strengthening and enhancing sea and inland dykes, rural roads, and canal maintenance (dredging) (SWIRP 2012). Maximising the potential of the Delta natural resources through using technologies and *modern* irrigation infrastructure continues to be in the Vietnamese government's agenda. The contemporary Delta farming system's high dependence on hydraulic infrastructure (mostly the canal system and dykes) and technologies ensures that the controlled water landscape with hydrocracies, intensive agricultural system and the *work-never-end* hydraulic investment will persist in the southern Delta.

A water-dependent livelihood: From adaptation to controlled water landscape

Throughout this history of building and reconstruction, the making of hydraulic infrastructure and irrigated agriculture have shaped and transformed Delta society, creating a rural life distinctive from the villages in the Red River Delta. The formation and transformation of the rural Mekong Delta has been neither a one-time formation nor completed change. Rather, rural Delta life and culture, associated with waterways, have been formed, adapted and reproduced throughout the history of human settlement, political changes and efforts to conquer nature. The case of rural Mekong Delta exemplifies the nature of institutions as *'arrangements between people which are reproduced and regularized across time and space and which are subject to constant processes of evolution and change'* (Cleaver 2012:8). This section reviews the transformation of rural Delta life that has been driven by the interactions among human settlements, social and political arrangements and nature.

Since the beginning of civilisation, inhabitants of the Mekong Delta have concluded that its '*deep river, flowing water*' are ideal land for human settlement. It has been believed that this kind of land has good Feng Shui (beneficial for prosperity), can prevent malaria, is convenient for transportation, offers water for drinking and irrigation and the possibility to dig wells in higher zones even in salty areas (Son Nam 1997:31). Interest in, and later traditional life based on the water has become not only a habitual, strong association with water resources and agriculture but also a cultural custom in the Delta.

The Delta's natural conditions, including the climate, flooding regimes and water landscape, have been not only the object of intervention but also the factors enabling and limiting the choice and effectiveness of interventions. Since the first settlement in the Delta, the water has posed risks (dangerous swamps) and opportunities (a water supply for everyday life in human settlements and rich aquatic fauna). Relying on the hydraulic water landscape throughout a long history of change into the present day, Delta life has continually seen water resources as playing a core role in all activities. The importance of water, rivers and canals has come to life in the Delta's literature, poems, songs and everyday conversations.

Many factors, particularly human interventions and the transformation of the local agriculture system, have formed and changed the water-reliant Delta. As asserted by Benedikter (2014b:12),

> Large-scale engineering campaigns did not only transform nature, but spatially rearranged human living environments such as settlement structures, the organisation of land use and access to natural resources in specific parcels of geographical space.

Human interventions to expand the hydraulic grid with more and larger scales of control structures (canals, dykes, pumps, sluices, culverts) has induced a shift from a traditional *river-water civilisation,* in which new settlers and their civilisation adapted to the natural conditions (Brocheux 1995:13); or the Mekong Delta 'amphibious' (ibid:3), towards a *modern hydraulic society*, in which water regulation and control serve as the key paradigms of a modern Delta civilisation that masters — not adapts to — the natural world (Benedikter 2014b:69, Evers and Benedikter 2009a, Evers and Benedikter 2009b, Pham Anh Tuan and Shannon 2011:103). In this *hydraulic society*, hydraulic management plays a predominant role in economic development (Evers and Benedikter 2009a) – the Delta's hydraulic agro-economy (term used by Benedikter 2014b:22). The present-day Delta consists of '*a man-made landscape comprised of a dense network of rivers, canals and ditches that crosscut an extremely flat terrain shaped by the omnipresence of rice fields stretching out to the horizon, only interrupted by houses, fruit orchards and arrays of trees planted along the water grid*' (Benedikter 2014b:69) and embankments and dykes acting as the barriers to floods and tidal inundation (Vo Khac Tri 2012: 74).

In addition to the hydraulic infrastructure, the political situation also plays a considerable role. Rulers have backed new technologies with policies and programmes designed to deliver benefits to certain segments of the farming community. Political struggles also affect the choice for farming system; for instance, wild rice cultivation returned during the escalating military struggles of the 1960s. The Delta's life, including its agriculture, has not been passively transformed; rather, the process of adapting to those elements has strengthened some human initiatives while subverting others (Biggs 2012:59). The survival of certain technologies, such as water pumps and dredges, illustrates that social recognition of their usefulness and the political situation supporting hydraulic farming with dredges are the primary influencing factors. How the social process and statecraft guide the practice and application of certain technologies and policies are analysed in the coming chapters.

Conclusions

The Mekong Delta is a prominent region for food production of Vietnam, playing a decisive role in the national food security and the export economy. The southern Delta has a diverse landscape officially and scientifically divided into zones for planning and management. The making of zones had been based on natural characteristics which determine the planning of and intervention in water infrastructure. Historical infrastructure investments have not only transformed the Delta's water landscape but have also formed and maintained the hydraulic agriculture that sustains the present-day national security and export economy and the formation and dynamics of the water-reliant livelihood. Throughout, there has been a persistent inter-relation between natural conditions and human interventions that has decided the choice of intervention and the effectiveness of investments. This inter-relation has guided the physical infrastructure and socio-economic situation of each zone and the areas within the Delta zones. The great differences between and within zones has created great diversity in the Delta, posing difficulties for the rulers to effectively implement new interventions, especially policies applied to large areas. In addition to constraints, the Delta's diversity also provides opportunities for human livelihoods. Thus, the natural conditions, with both opportunities and constraints, have helped shape the contemporary Delta water landscape.

Complex history of the making of the Mekong Delta's water landscape: Work in progress and continuity over history

The diverse nature of Delta and its social culture and political complexity, derived from a complex history of human migration, colonisation and nation-building, have been the main factors determining the characteristics of the contemporary Delta. The Delta has a long history of human settlement and intervention in the water landscape (Figure 3-5), starting around 2,000 years ago with the first recorded civilisation of the Funan people, followed by the Chenla and, later, the Kingdom of Khmer. The pre-Vietnamese civilisations made the first human interventions in the Delta, using natural resources to meet various needs. While constructing several basic hydraulic infrastructures such as canals, the first settlers tended to adapt to nature, rather than control it. Even before the southward migration of the Vietnamese to the Delta, its inhabitants left imprints on the water landscape. Especially during the Khmer occupation, innovations in rice varieties and cultivation techniques enabled '*bringing water to the field*' through canal projects, marking the first form of irrigated or hydraulic agriculture in the Delta. The water infrastructure was then expanded with canals, forts, roads, and markets through investments by the Khmer and the Chinese. Those structures formed the foundation for the subsequent investments by the Vietnamese who moved south during the late 1700s and the organised South migration in the Nguyen era during the early 1880s. During the Nguyen era, enormous investments served mostly security and military purposes to bolster the dynasty's claim to the territory. In addition, the farming intensification, aligned with canal projects, was at the centre of management from the 1830s, particularly in land management and taxation policies.

Canal projects were continued and restored during the first colonial period under the French. This period was characterised by mechanised canal construction using machines, instead of soldiers or conscripted labour. The colonial administration was led by engineers and construction companies who, through the regulating role of machines and *modern* technologies, held monopoly power over projects. Despite the continuity in prioritising the military and security targets, French colonial hydraulic investments also supported agricultural production through both official and individual-initiated projects. Overall, politicised hydraulic interventions remained dominant in both this period and the balkanisation period, which resulted in 2 separate zones under different water intervention regime. Later, during the American war, the existing water landscape continued to serve as the base for further investments in canal system to mechanise and modernise the Delta's agriculture. In addition to wartime destruction, the American colonial period saw the introduction of significant agricultural technologies, including water pumps which remain in wide use in the rural Mekong Delta.

The post-colonial period can be described as the most important influence on the present-day hydraulic water landscape of the Delta. Needing to secure food production and develop the economy as an independent country, the single investor — the Vietnamese state — has more reason and capacity to prioritise investment in hydraulic infrastructure to sustain agriculture production, rather than solely to serve military or security purposes (which remained the priority in the 1970s and 1980s after the reunification). The nation-building programme in this period, which has been influenced and supported by international ideologies and donors, has expanded the canal system and other hydraulic structures essential to Delta's intensive irrigated agriculture and transportation system. The programme has also maintained existing canals and invested in dykes, embankments and pumping stations for flood mitigation.

Figure 3-5: Overview of the history of the making of the water landscape of the Mekong Delta (designed by author)

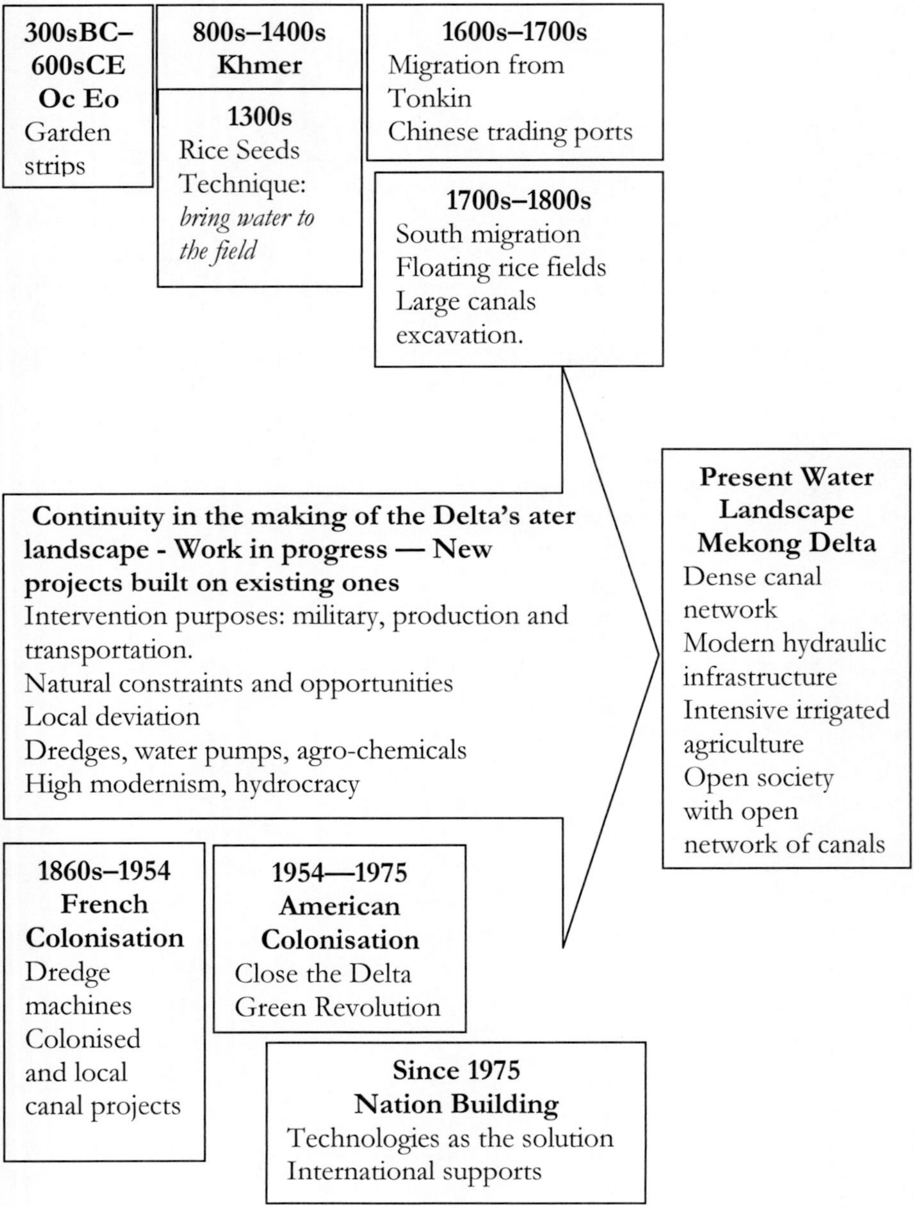

The history of interventions in the Delta's water landscape reflects the *work in progress*, with new infrastructure built on the old and continuity maintained as many past investments and innovations remain either full or partly in the future landscape. The *work in progress* continues as the Delta's people want to use water resources for various activities, ranging from irrigation and aquaculture to transportation, domestic consumption and recreation. History illustrates that the selected interventions and planed water landscape depend very much on the natural conditions, pre-existing infrastructure, social selection of policies and technologies and political situation, in which the rulers are the most influenced planners. Nature gives human settlements and their interventions powerful opportunities (rich water resources, fertile land) and constraints (*sacred forests, poisonous water,* swamps, season flooding, dead points blocking canals with sedimentation). The available infrastructure supplied the foundation for subsequent interventions, although many mistakes made in the attempts to conquer the nature have persisted across periods of time. The social selection of policies and technologies has relied on the existing culture and political situation to decide what to keep and what to demolish.

Amongst many others, the political factors, including the ruler's role and the political situation, have a decisive role in building the Delta's water landscape. The ruler or the state has planned most large projects in either *opening* (from the beginning of human intervention) or *closing* (shifting to a hydraulic water landscape with closed structures) the Delta water network. Their intentions can target various goals whose priority varies by period. In general, during the pre-colonial, colonial and post-colonial periods, hydraulic projects have been mostly by military and security purposes and occasionally by economic and agriculture interests, which have become more prominent in the post-colonial period. With the investor's power to set administrative structure and budget, the ruler or state has played the primary role in shaping the Delta's water landscape.

Rulers have generally used hydraulic and other technologies as tools to control territory and to ensure and increase the economic and social benefits of certain groups (e.g. the case of dredging companies with dredging machines under the French colonisation). The tools themselves become an interface for interaction and negotiation among different individuals and groups. The tendency to maintain control through technologies or the ideology of modernism continues in the independent state regime. Benedikter (2014b:60) concludes that water regulation has not only become the key paradigm for achieving growth, prosperity and modernity but has also played a strategic role in supporting the legitimisation of the state and the establishment of new rural institutions. Thus, in order to maintain the hydraulic water landscape for multiple purposes, the present-day Vietnamese state has continued to invest in hydraulic infrastructure, despite the softening of scientific consensus on exclusively technocratic hydraulic interventions.

However, regardless of desire, no precolonial, colonial or postcolonial ruler has achieved complete control over territory, nature and society. While French colonial authorities and the 1970s–1980s Northern Vietnamese bureaucrats were unfamiliar with the Delta's nature and faced natural and social challenges in implementing their policies, other rulers, including the contemporary Vietnamese state, have not ever effectively addressed the Delta's diversity in natural and physical conditions and local dynamics. Although nearly all of the Delta's surface bears signs of human interventions (Biggs 2012:6), that does not grant complete control over nature. Settlers in the Delta have encountered both benefits and constraints from nature. Consequently, all have learned to apply trial-and-error approach in order to control and/or adapt to nature. In addition to natural constraints, the Delta's rulers have also faced various responses from the settlers who sometimes react differently to authorities' plans because they have different backgrounds, cultures and advantages. Specifically, as seen through history, the introduced technologies have been subjected to social selection which decides the survival of the technologies. In previous eras, local responses have been expressed in various forms, ranging from covert to overt resistance, from silently and strategically modifying the ruler's mandate to open unrest.

Recognising these local dynamics, the contemporary Vietnamese state seems to take a more strategic approach to dividing the levels and extent of interference. Using hydraulic structure to determine the threshold for permitted local flexibility and mandatory state control, the state uses its presence as a strategic solution to achieve the dual goals of state control and adaptation to local dynamics. These issues are elaborated in the analyses of the state structure in Chapter 4 and the implementation of state policy in Chapter 5.

Institutions as the outcome of continuous evolution throughout history

Each historical period of the making of the Delta's water landscape has seen attempts to impose foreign rules and arrangements and sometimes to replace the existing system. However, pre-existing elements have survived or been transformed through a process of interaction and negotiation in which actors and stakeholders influence others and join in selecting the current institutions. Consequently, the Delta culture, agriculture and rural livelihood have been formed from the history of interaction of various institutions. The transformation of agriculture and rural livelihood is not a straightforward process. Rather, practices have been formed, negotiated and adapted through interaction between rulers' initiation of interventions and local inhabitants' selection of intervention, all under the influence of natural and physical conditions.

Institutions connect the past and present, old and new, traditional or existing and imported arrangements. According to Cleaver (2012:8), working from the school of critical institutionalism, the complexity of institutions is intertwined with social life, with their historical formation under social and economic changes over time and with the interplay between global and local factors, the traditional and the modern, formal and informal arrangements. The interaction of old and new and central and local institutions occurs throughout the history of the making of the water landscape the Mekong Delta. As Jamieson (1995:2) states, '*all Vietnamese people are today still, as they were fifty years ago, interacting with that past in the process of shaping their future*'. In the field of hydraulic management, Biggs et al. (2009:221) argues that the reproduction of past arrangements reliant on private contractors to carry out public works since the 1880s has continued to shape state decisions because of the interests of project lobbyists and the aging of infrastructure, which leaves no space for major changes in water resources strategies. Biggs (2012) terms this continuity institutional and physical inertia or the powerful influence of pass arrangements which persist either because the reasons for their existence remain or because popular belief holds that certain ideologies (e.g. the technocratic solution of building hydraulic works) is the best solution (e.g. fast, easy, more beneficial) for development.

However, institutional and physical inertia have not been able to preserve any single institutional or physical setting. Thus, changes and amendments have happened in the formation of bricolage (Cleaver 2012; see Chapter 2 for more background information). Institutional bricolage has been formed through interventions throughout the pre-colonial, colonial and post-colonial periods. Institutional bricolage makes up modern hydraulic agriculture, with a canal irrigation system started in the early settlement of the Delta, dredging machines introduced by the French and individual pumps first brought to the Delta by Americans. The whole system and policies continue to be heavily dependent on technologies and infrastructure (institutional inertia) but pay a certain level of attention to local flexibility as a result of the adaptation process over the history of state management. The same tendency can be seen in the Red River Delta, where water control consists of a mixture of features from the collectivist period and new constructs from 2 decades of liberalisation and decentralisation (Fontenelle 2000).

In addition, the living and working culture of the Southern Delta reflects various cultures involved in the complex history of interaction with nature and human settlements. Sox (1972:140) concludes that Vietnamese adaptations and mal-adaptations are a reconstruction of the Cham exploitative systems (e.g. crop cycles, irrigation subsystems, boating, fishing and some economic organisations). The interrelations of Khmer peasants, fishermen and wood cutters, Chinese merchant-lenders and transporters and, later, Confucianism have made the pluralistic Delta society which shares economic tasks and emphasises paternalism (Brocheux 1995:108, 112, Hickey 1968:5, 134). The complexity of changes and various cultures from different groups and periods has created the diverse society of the South. The experience of diversity over time has made an open society that easily accepts new factors entering the institutional negotiation ground. '*The villagers feel that the old ways are good, especially when they are expressed in rituals, but they also feel that the new ways are acceptable when proven of value*' (research in the Mekong Delta, Hickey 1968:285). As follows, Delta inhabitants have assessed and validated new technologies and practices, including irrigated farming and the usage of individual pumps. This behaviour marks the process of legitimating institutions, or qualifying pieces to contribute to a new bricolage.

The interrelations among natural, physical, social and political conditions generally have been the main drivers of water resources management over the history of making of the Delta's water landscape. The contemporary state shares with the pre-colonial and colonial governments the desire to control territory. The question then is how the state achieves its goal in the contemporary period in which the market and individualistic lifestyles are increasingly dominant in the Delta. The analysis of the bureaucratic–informal interface of water management practices in the Mekong Delta presented in the following chapters sheds light on these issues. Chapter 4 introduces and analyses the state management structure which affects internal dynamics, especially at the commune–hamlet level, and other actors at the interface. Chapter 5 presents an anthropological analysis of the negotiation at the interface over water drainage and canal dredging. We point to the role of the institutional process in guiding and, thus, explaining the *work in progress* in Delta's hydraulic intervention and the influence, survival and transformations of innovations in the Delta landscape over history. Chapter 5 contributes to the understanding of the role of the institutional process by illustrating the process of institutional legitimation and by partly answering the question of how bricolage is formed in contemporary irrigation management. In analysing the activities of field drainage and canal dredging, the negotiation processes are reviewed. Examining the processes involved in the power relations and interrelations between the state and other actors clarifies the nature of state–society interaction in the form of everyday dialogue in the Delta's local irrigation management.

CHAPTER 4

DECENTRALISATION AND IRRIGATION MANAGEMENT IN THE MEKONG DELTA

In the hydroscape of the Mekong Delta, the water resources and hydraulic infrastructure, essential for irrigated agriculture and human livelihood, are increasingly dependent on human interventions. In Vietnam, as in many other countries with state-centric governance, the state plays an important role in directing, monitoring and facilitating socio-economic conditions and the management of various sectors, including the water sector. The importance of the role of the rulers or the state in the making of the Mekong Delta's hydraulic water landscape has been proven historically, as analysed in Chapter 3. However, the state neither acts alone nor has the primary decisive power in all matters. Rather, practices are guided by complex, existing social–political processes in which various stakeholders, strongly or partly tied to the state system, interact with each other at different interfaces.

As one stakeholder, the contemporary Vietnamese state manages the nation, including agriculture and irrigation, with a Vietnamese version of decentralisation. In this system, power, roles and responsibilities are divided among administrative levels. Decentralisation in Vietnam was initiated in response to social pressures emerging from social and cultural diversity. As well, the political and economic shift to a market orientation gave the state the opportunity to retreat in face of increasing financial and regulatory burdens. This retreat presaged the top-down decentralisation process inscribed in the legislation.

This work suggests that the Vietnamese state structure has no clear divisions that one can objectively outline. Instead, macro divisions embedded in legislation create and allow a framework for enforcement and implementation at the lower levels. Analysis of the local levels reveals that official decentralisation has taken place in parallel with unofficial decentralisation. The latter often has led or preceded the former in guiding execution in localities. This chapter sheds the light on how decentralisation has manifested in the Mekong Delta and affects the management of the irrigation sector.

I argue that the pattern of decentralisation with looser control over local governments in finance and decision making, complemented by the parallel processes of privatisation and localisation arising from commercialisation, is a response to the decline in the state's capacity to apply the command-and-control management approach. That said, I do not claim that decentralisation eliminates the state's status at the local level; rather, the state remains present in various material and symbolic ways. To a certain extent, the state continues to intervene, facilitate, encourage and direct socio-economic processes in order to achieve national goals and avoid political risk. In contrast to the command-and-control regime, especially in its extreme manifestation during collectivisation in the 1970s–1980s, the contemporary Delta agriculture and irrigation systems indicate a flexible system, relying on local dynamics to manage local affairs while reserving a guiding and monitoring role for the state. By analysing the decentralised state structure for irrigation and agriculture and outlining the relevant stakeholders, this chapter serves as a starting point for the analysis on local dynamics in irrigation management presented in Chapter 5.

This chapter consists of 2 parts. The first analyses the decentralised structure of state management, including in the irrigation sector. The institutional set-up and the management mechanisms, including budget management, that reproduce task division among the 4 state levels are then elaborated. The institutional set-up from the provincial level downward is analysed in the case of Can Tho, the main research site. The second part describes the individuals, agencies and organisations involved in irrigation management, their ties to the state, their functions and their general influences on management of the sector. This analysis complements previous studies on the Mekong Delta, especially the ones from the WISDOM project[35] which have investigated, among other topics, the general state management of the water sector (Waibel 2010) and water management at the planning level, including policy-making processes from the central to the provincial level (Pham Cong Huu 2011, Reis 2010). The question which remains is how policy is actually implemented, interpreted and used in practice, given the diversity in the natural, physical and socio-economic conditions of the Delta and involvement of varied stakeholders. To get closer to a level of analysis centred on individuals and organisations, this chapter focuses on the local level, specifically the district, commune and hamlet levels.

Decentralisation in irrigation management

Superficially, the irrigation sector is organised through a decentralised system composed of individuals and agencies across the 4 administrative levels. Of

[35] WISDOM is a joint Vietnamese and German project that has designed and implemented an information system in the Mekong Delta to capture information from the fields of hydrology, sociology, information technology and earth observation. http://www.wisdom.caf.dlr.de/en

these, fiscal arrangements appear to be the main target in promoting decentralisation. In this realm, the provincial level is responsible for service provision and infrastructure maintenance, notably the repair of canals and dykes (since 1986) (Biggs et al. 2009:207). However, Fritzen (2002) criticises that decentralisation in Vietnam as incomplete, resulting in a de-concentrated structure from the central to the provincial level and a strictly centralised structure from the provincial to the local level. Additionally, regardless of the characteristics and extent of decentralisation at different levels, the distance between state policies and real practices depends significantly on local dynamics created by various stakeholders through interaction and negotiation.

This section first reviews Vietnam's hierarchical administrative system of state agencies from the central to the local level. Aspects of local autonomy within the state management system, especially in the context of the Mekong Delta, are described. The existing challenges and issues of decentralisation in the irrigation sector are analysed through an empirical account of budget management for canal dredging in Can Tho. Finally, the inconsistent state budget system that allows local cadres flexibility to acquiring many resources to fulfil structural shortages is discussed.

Vietnam's administrative structure

Vietnam's administration system has a top-down, 4-level organisational structure: the central or national level, provincial level, district level and commune level. Officially, the commune is the smallest unit of the state administrational system (Porter 1993). However, below commune, there are hamlets with People's Board, cadres or state assistants tasked with assisting the commune. In general, the central level includes the National Assembly, ministries (including the Prime minister's office and various departments) and the Supreme Court. These agencies fulfil the legislative, executive, and judicial functions of the national government, respectively (ibid:73–83). At the province and district levels, the People's Council, People's Committee, and People's Court perform the legislative, executive and judicial functions, respectively. Within the ministries exist 3 types of agencies: state management, non-business (generally translated as institutes), and business agencies. The latter 2 are governed primarily by the state management agencies (Molle and Hoanh 2008, cited in Waibel 2010:18).

Ideologically, authority to run the nation-state in Vietnam goes beyond governmental institutions. A popular slogan promoted in Vietnam asserts: '*The Party leads, the People controls, and the State manages*'. As stated, the nation is managed by the state under the Communist Party[36], with the support of People's Army, Fatherland Front and mass organisations (e.g. for women, peasants, workers, youth) (Kerkvliet and Marr 2004:3-4). As at the state system, the Communist Party is present in all levels and form of office (province, district) and individual agents (commune and hamlet).

Official speeches and state cadres' conversations indicates a separation between the central level (*trung ương*) and the local level (*cấp địa phương*), in which the latter views the state system from the province downward. The local level is referred to as *chính quyền địa phương,* which, depending on the context, can be used to describe the local government, local administration or local authority (Kerkvliet and Marr 2004:3). In this research, we use the term 'local level' to refer to the commune which consists of hamlets, although in certain, specified cases, this research combines the district level with the local level. At the commune level, the governmental and party structures often overlap, with cadres holding positions in both systems. At the lowest level of the state administration structure, the Party and state join in implementing policies and arranging the local practices. Therefore, differently than the slogan, the Party leads, the state executes, local arrangements become contextualised so that, depending on the local state structure, the actual manager of the local state–party system for activities can be a state cadre, a Party cadre or a double-tasking cadre. During interviews, local cadres stated that the local state assists in implementing policies given as mandates from higher levels. In other words, the local level is the 'extended' arm of the higher hierarchical level (interview 29.08.2011, 14.12.2011).

The "modern" 4-level administrative structure, from its nomenclature to its structure, has been established, copied and transformed over Vietnamese history. This set-up of administrative levels was practiced in Tonkin in Northern Vietnam under the influence of Chinese system from around 100 BCE until the 10th century CE. Vietnamese society then was based on Confucian and, later, neo-Confucian ideologies (Marr 2004:28–30). This traditional division of state structure was maintained and adjusted by various pre-colonial regimes. Marr (2004) asserts that, as far back as the 15th century under the Le dynasty (1428–1788), the territory was divided into regions (*xứ*), standing over prefectures (*phủ*), sub-prefectures (*châu*), districts, communes and other units with diverse designations (ibid:29). The term for commune, *xã*, was

[36] It should be noted that a significant percentage of state cadres are Party members. For instance, nearly all the presidents of People's Committees are prominent members of the respected local branch of the Communist party (Kerkvliet 2004:9).

first used in this period to designate '*an organisational template to be applied to every existing village, making them the bottom rung of the state ladder*' (Marr 2004:30).

The colonial rulers continued imposing a hierarchical structure during their management and control of the colony. During the 1860s, when the French first arrived in Vietnam, they established a control structure down to the canton level (*tổng*, one level above commune) to implement taxation, landownership, justices, and education policies (Marr 2004:35). The present organisational structure was established in the Northern Vietnam after 1946 and copied in the South after the reunification in 1975 (Kerkvliet and Marr 2004:4). Thus, the division of tasks in the governmental system has been copied and adjusted throughout a complex history of nation building, colonisation and independence. However, despite its roots in earlier regimes, the hierarchical system has had a significant presence at the commune and village level only since French colonisation, and only a certain degree of de-concentration took place after Doi Moi with the decentralisation policy.

Vietnam's present-day bureaucracy is principally organised according the principle of double subordination. At the province, district and commune levels, a People's Committee runs day-to-day state affairs, such as implementing state policies and supervising and giving mandates to the branch offices of national ministries and departments. The People's Committee acts as the executive arm of the People's Council at the same administrative level. Simultaneously, though, the Committee is accountable to the People's Committee of the level above. Following this pattern of subordination, the branch offices of ministries report to their line-offices at the next highest level. The vertical and horizontal accountability in all agencies in Vietnam is called double subordination (Kerkvliet and Marr 2004:8). This dual subordination structure is, in effect, a strategy to maintain control over the nation, to encourage and monitor the implementation of policies and state targets (vertical accountability) and to coordinate assessment at one level (horizontal accountability).

Waibel (2010:18) describes the working structure of the state:

> State management functions are implemented by administrative and professional departments and include policy making, advisory functions, the organisation and monitoring of the implementation of laws and policies and administrative and other tasks, such as international cooperation. Responsibilities, tasks and functions of each ministry are defined by individual decrees.

In general, the bureaucratic system is regulated by legal sets. the Vietnamese legal framework has many constraints because of consistency among documents, as illustrated by Nguyen Thi Phuong Loan (2010b) in the case of water sector. However, Waibel (2010:15) contends that, even in the top-down approach to decision making, policy formation and communication, many policies remain floating; consequently, their implementation and the state capacity to control them are limited. In addition, policy implementation at the local level is subject to deviation due to local flexibility. Waibel (2010:15) explained that local flexibility occurs because officials at the commune level '*remain in their home communities and often assume local state management tasks only in addition to their farming activities*'. Consequently, '*local government institutions are both part of the state management system and bound to local interests and conditions*' (ibid). The co-existence of centralised policy-making and local flexibility has invariably been assessed as a technical weakness of the state (Gainsborough 2005:48); for instance, it has been argued that '*the machinery is incompetent and often responds more to local or personal interests than to what the party-state says or law dictates*' (Koh 2001:536). However, the state's status and the relations between the central and the local levels are the result of a more complex process in which individuals and groups interact in possible negotiations and conflict. Before analysing that process in the contemporary period, which is the main target of this work, the following section sheds the light on the history of local–central relations which has been described with the common Vietnamese proverb '*The King's laws give into the village's norms*'.

The historical central–local relation: 'The King's laws give into the village's norms'

Historically, village autonomy has frequently presented itself as a target for intervention and as a headache for the ruling monarchist or the state. In the pre-colonial period, the ruling elites' principle of demarcating smaller territorial units and assigning precise responsibilities, in theory, was applied consistently across the land, but '*officials posted more than a hundred kilometres or so from the royal capital possessed considerable administrative discretion*' (Marr 2004:29,46). This suggests that, in a way, during the monarchist era, Vietnamese villages enjoyed a certain level of '*democracy*' or '*autonomy*' (Marr 2004:31-32). The tendency was somewhat continued and reached a comparable degree of autonomy from central administrative control in the second half of the 10th century, a governance system which the French later labelled as '*small republics*' (Grossheim 2004:54). In the subsequent period of colonisation, all types of authorities struggled to exert full control over local villages, resulting in considerable efforts combined with instability. Grossheim (2004:74) asserts that the French did not gain power over local authorities because they undermined the local traditional arrangements with favouring powerful old notables. Those notables had maintained a level of autonomy in managing the village affairs since the pre-colonial period (ibid:57–58). Today, in contemporary Vietnam, Marr (2004:47) comments that the Communist party–state has '*succeeded in penetrating the village to a degree impossible under former rulers*'; nevertheless, the impacts of the current administrative structure on local government remains unclear.

Amid the market economy and social changes, it asked how autonomous from or dependent on the central level the local authority needs to be or should be allowed to be. In everyday conversations, '*Phep vua thua le lang*' ('*The King's laws give in to the village's norms*') remains a common proverb, widely used by local cadres (communes and hamlets) during this research. The statement expresses a popular view of how local people, guided by local cadres (or former cadres or prestige village elites), can create slippage in the local internal affairs for the benefit of the locality. Indeed, it is stated in the research area that, according to the local conditions, '*the commune level* [commune state] *has the decisive role. The district level only gives instruction, monitors the implementation*' (district cadre 14.04.2012). The statement from a district official illustrates the acceptance at higher hierarchical level of some autonomy at the local level, which underlines the continuity of meaning of the proverbs about central–local relations in Vietnam.

However, repeating the proverb could be habitual and not necessarily convey the same meaning compared as in the past. Its meaning changes slightly in locations across Vietnam. Indeed, local interests are not consistently put above national state interests; it depends on the suitability of the policy, the allowances of command to local cadres given by the present political regime, the local dynamics and background in which the local officials operate, and their knowledge of and prestige among the people. Regional cultural differences also influence local response. For instance, after a long period of rule by the command-and-control regime, local government agencies in Red River Delta had more incentive to stick to the mandate for collectivisation than the Mekong Delta, which had had an open, capitalist market under the Americans. Grossheim (2004:56) states that Mekong Delta society differs significantly from the traditional Northern Vietnamese villages because '*they were much more 'open' and lacked the strong community institutions like the public land system that characterized villages of the north*'. In addition, the open social culture of the southern Delta (see Chapter 2) can offer more opportunities for adaptation and acceptance of new things, possibly leading to a higher degree of compliance with the state's policy of technologies transfer than in the more closed villages of the North. Overall, deciding between official policies and local interests involves a process of negotiation and accomodation. Thus depending on the policy, locality and time, local response can be compliance, deviation or opposition to national policies.

Thus, we arrive at the question of how the state positions itself in a context such as the Mekong Delta. As contemporary Vietnam moves towards a market-influenced economy and inconstantly changes state legislation and management structure, it becomes necessary to re-characterise the state, looking deeper at the degree to which it fits labels such as weak, dominating, responsive, or repressive. This task is done through an anthropological analysis of state management at the local level in Chapter 5. As a background illustrative of several structural dynamics of state power, this chapter presents an analysis of state management in irrigation sector and the remaining challenges facing the decentralised and decentralising system. The last section reviews stakeholders joining the practice of negotiation and analyses their position within the interaction space and in relation to the state system.

Decentralised structure of state management in the irrigation sector

Since Doi Moi was officially launched in the 6th Communist Party Congress in 1986, Vietnam has undergone economy liberalisation aimed at '*developing the market economy with multiple components according to the market mechanism under state management and oriented towards socialism*' (Article 15, Chapter II, Constitution 1992). During the 1990s, state policies focused on the decentralisation of state management in order to improve the effectiveness of national policies and ensure that they reach the lowest level: the people. The Vietnamese state adopted a statement by President Ho Chi Minh — '*the commune level, being closest to the people, is the foundation of our public administration. If the commune level works, then all our work will proceed smoothly*' (translated in Fritzen 2002:2[37]) — as the core principle of its decentralisation policy.

Fritzen (2002:9) asserts that Vietnam's state's decentralisation consists of 2 elements. First is de-concentration, in which the central level assigns specific implementation duties to different levels. Second is political devolution, in which '*local governments acquire real discretion, resources and, as the phrase implies, rights*'. (ibid:9). Decentralisation has been implemented in Vietnam in 3 dimensions: fiscal, administrative and political. While fiscal and administrative decentralisation has been carried out through administrative reforms (*cải cách hành chính*), political decentralisation is conducted through grassroots democratisation (*dân chủ hóa cơ sở*) with the aim of giving more power to local levels down to commune level (Ordinance on Grassroots Democracy —

[37] Translation by the author of the following: '*Cấp xã là gần gũi nhân dân nhất, là nền tảng của hành chính. Cấp xã làm được việc thì mọi công việc đề xong xuôi*', *Hồ Chí Minh Toàn Tập*, Volume 5, p. 371. Nhà Xuất Bản Chính Trị Quốc Gia, 1995.

Practicing Democracy in Commune, Ward and Town 34/2007/PL-UBTVQH11[38]).

As it enters the 21th century, Vietnam has increased participation by private entities in various state-managed activities, following socialisation and privatisation policies since the early 2000s. The socialisation (*xã hội hóa*) policy as stipulated in the Socio-Economic Development Plan 2006–2010, by Norlund et al. (2007:6), calls for society to make contributions to education, vocational training, health, culture, sport and the environment. This policy has opened up the country's economy to private sector participation through privatisation in the form of equitisation (*cổ phần hóa*).

> Equitisation is defined as the transformation of SOEs into joint-stock companies and selling part of the shares in the company to private investors in order to improve the performance of the firms in question. (Truong Dong Loc 2006:41)

Similarly, the newly equitised firms exist in a space between state control and private sector. Gainsborough (2009:266, 268) describes equitisation in Vietnam as *private indirect government*, a notion which Hibou (2004) uses to contend that the power over enterprises the state has retained as a shareholders is uncertainty. Among others, many state cadres operate private enterprises[39].

Water resource management at national level is the responsibility of different ministries, depending on their activities (see Nguyen Thi Phuong Loan 2010a:62 for a chart of state responsibilities for water resources management at national level). Since the MONRE was established in 2002 (Government Decree 91/2002/NĐ-CP), there has been an on-going rearrangement of tasks between MONRE and the MARD, the 2 primary ministries managing the water sector. Increasingly, tasks have been transferred from MARD to MONRE and its line agencies. The water-related duties of MONRE and MARD can be summarised as follows:

- MONRE: responsible for general water resources, mineral resources, geology, the environment, meteorology, hydrology, cartography, sea and inlands

---

[38] This replaced the Grassroots Democracy Decree 79/2003/NĐ-CP, whose main theme was '*people know, people discuss, people do and people inspect*'.

[39] This was confirmed by Evers and Benedikter (2009) based on the cases of private hydraulics construction firms in the Mekong Delta.

- MARD: manages agriculture, forestry, salt-making, fisheries, irrigation, flood and disaster prevention and rural development

In the incomplete transfer of tasks between the 2 ministries, irrigation remains under the purview of MARD because it is closely related to agriculture, which falls under MARD's management umbrella. There has been competition and confusion about the designation of authority and responsibility for various water activities between MARD and MONRE (Benedikter 2014b:148-149). Ultimately, the hydrocracies rely on irrigation to consolidate state power, and as long as they maintain the existing state structure for irrigation management, the 2 different ministries are unlikely to dilute this process.

Today, irrigation is basically managed by MARD, its line offices and cadres across various administrative levels[40] (see Figure 4-6 for an example of the state organisational set-up of irrigation management from Can Tho). MARD includes the Department of Agriculture and Rural Development at the provincial level (DARD) and the Offices of DARD at the district level, which are responsible for the general management tasks in irrigation. DARD's district offices are called either offices of agriculture for the rural district (*huyện*) or offices of economics for urban district (*quận*). In the transitional Can Tho, half of urban districts still have agriculture as their main economic activities. Thus, the offices of both economics and agriculture are responsible for implementing agricultural mandates from DARD.

The duties of the DARD district offices include planning agricultural production; implementing the sowing schedule for rice cultivation; providing technical support for communes in co-operation with other professional agencies, such as extension offices, irrigation stations and plant protection collaborators; facilitating the mechanisation of agriculture sector to support commune cadres in arranging harvest contracting between farmers and companies; and receiving reports from the communes about the production situation and communicating the reports to DARD. In irrigation maintenance, the DARD district offices either assist the provincial sub-departments of irrigation in managing dredging projects or directly organise smaller dredging projects. Reports about the condition of canals in the district to the provincial level are one product of these activities. DARD district offices also take responsibility for implementing other national programmes, such as the

[40] See Decree No. 01/2008/ND-CP for the detail of tasks assignment to MARD and DARD towards water resources management focusing on irrigation (Bucx et al. 2010:32).

standard field[41] and new rural area[42] programmes. As seen in all the districts I observed (3 agricultural districts and 1 in transition to an industrial urban district), the DARD's office is the most important and leading office at the district level wherever agriculture remains an locality's main income source. Thus, the office wields significant power over the arrangements of irrigation and agriculture production and irrigation maintenance. Based on its position at the border between policy and local state arrangements, it repackages existing accomodations at the commune level into reports that suit the demands of provincial and national ministries.

In addition to the generalised tasks of irrigation management, peripheral duties, such as providing technologies and directing specific construction, operation and maintenance projects, are delegates to professional agencies: Department of Water Resource Management at the ministerial level and its branches. The Sub-Department of Irrigation Management performs these tasks at the provincial level, irrigation stations at the district level, and irrigation and rural transportation cadre at the commune level. At the hamlet level, the irrigation sector falls within hamlet affairs and is taken care by hamlet People's Board. While the irrigation offices at the central, provincial and district levels take on the general responsibility for managing irrigation, irrigation cadres in the communes can have a wider scope of work, depending on local arrangements. In most cases, irrigation and rural road cadres have the leading role in organising agricultural production, including arranging irrigation and drainage activities, facilitating the contract process and implementing various state programmes with other commune cadres, assisting and/or organising canal dredging projects and managing rural road construction, maintenance and repairs. Additionally, extension cadres and plant protection collaborators often based in the commune are also active in irrigation activities within the scope of their responsibilities for agricultural production. According to horizontal

[41] Standard field (*Cánh đồng mẫu)* is an agricultural program which encourages consolidating fields to cultivate the same variety at the same time, along with jointly managing irrigation with a closed dyke system, mechanising all stages of cultivation, and making sales contract and/or farming contracts with the business sector. This model started in An Giang during the 2010 winter–spring season and has spreading in the delta and other areas in Vietnam.

[42] The building new rural area (*Nông thôn mới*) programme aims to construct and organise the life of rural residents in a modern, civilised way and to preserve culture and the environment. This programme is intended to support infrastructure for social and economic development suitable for local planning and to combine support from the State in enhancing internal capacity (Nguyen Anh Thuy 2011).

subordination, the professional agencies are under the supervision of and accountable to the general management office at the same level: the People's Committees. Vertically, they have report to their line agencies one level higher. Thus, the organisations of irrigation management and agricultural production have many points of contact, creating an environment permitting adjustment and adaptation to the local situation.

This research in Can Tho illustrates that the actual task and efficiency of irrigation cadres varies among communes, depending on the incentive to work effectively, which increases individuals' prestige with commune's leaders and leads to '*being trusted to take more responsibilities*' (interview 18.10.2011). For example, at the commune level, instead of the hierarchical division of task regulated by legal documents, task division is flexible, driven by kinship, personal relations (good at working with others) and past personal performance. This phenomenon is further analysed in Chapter 5, focusing on a case of sowing and field drainage management in which the localised task division helps increase the commune's working efficiency but also reflects inequality in work, which might allow laziness on the part of some cadres and space for others to hold a power monopoly. In many cases, the local arrangements for task division aim to fulfil or balance multiple purposes or, in other words, multiple pressures. One cadre might want to fulfil both state mandates and personal interests, perhaps gaining prestige to climb to a higher state position or financial or other in-kind benefits through projects and programmes (officially or informally). Other cadres might also consider the community's demands that are not recognised or contrary to state mandates.

Actual arrangements of who does what often comes after formality. At the commune level, formality is expressed in the annual establishment of Agricultural Production Steering Board during the meeting for production planning before the winter–spring rice season. The boards consists of hamlet leaders, along with the general managers of the commune, the People's Committee leaders and professional, extension, irrigation and rural transportation cadres. As well, the leaders of the commune's Party unit can serve as the Production Board's leader, if the position is not given to a leader of the People's Committee. With this task division among the leaders, only 1 of the Party's leaders and 1 of the People's Committee's leaders join the Production Board. For the People's Committee, the vice president responsible for economy (the other is responsible for social and cultural activities) receives the responsibility. Frequently, 1 cadre holds a double position in both the People's Committee and the commune's Party unit. The last seats on the Production Board go to representatives of mass organisations, mostly from farmers', women's, veteran's and youth's unions.

The gathering of state and party cadres responsible for the socio-economic fields in the Production Board illustrates the state's attempts to enhance the collaboration of local cadres in a united force implementing state policy at the commune level. This system can be considered as corresponding institutional reform, as contended by Waibel (2010:22). The Production Board members are tasked with guiding, managing and reporting the hamlet's and commune's production. The board is especially active in the winter–spring season, when rice planting over a large area and field drainage before sowing, which requires much organisational effort in many parts of the Delta, are planned. In general, each leader of the Commune People's Committee (CPC), Party unit and mass organisations is appointed to a directing role in 1 or 2 hamlets. These cadres are responsible for supervising all affairs in these hamlets and reporting them to the commune. This management structure of the Production Board illustrates the ideology that has guided the Vietnamese party–state working system: '*collective leadership, individual responsibility*'. This ideology is adapted by the party–state in recognition of Ho Chi Minh's idea of management.

> One person, despite talent and experience, only recognises and considers one or some of the many aspects of an issue but not all. Therefore, we need more people to provide more experience; some can see this aspect, while others recognise other aspects of the same issue. With contributions of experience and opinions on the issue from different people, the issue or problem is assessed thoroughly. Only when we see all the aspects will problems be solved effectively without any mistake. (Ho Chi Minh, cited in Ho Chi Minh toan tap 2011:619, cited in Nguyen The Thang 2012, translated by the author)

However, the practices at the commune rely more on local arrangements of leadership and decision making. Most of these arrangements do not involve all cadres supposedly involved in the collective leadership, and there is consistent overlap of responsibility. Sometimes, the gathering of cadres into the Production Board is a symbolic bureaucratic task as only 2-3 cadres actually carry out the production planning tasks. In the 3 researched communes, 1 or more of the group of 3 — the irrigation and rural road cadre, the extension cadre and the Farmer's Union cadre — served as the main managers and facilitators responsible for commune's production. The task division differs among communes depending on their specific arrangement. However, the lack of an official collective meeting for decision making does not necessarily mean that the absence of collective leadership. At the local level, collective leadership is carried out in a more flexible form through informal and individual communications. The local forms of decision making are illustrated in Chapter 5 through the management of field drainage for the 2011–2012 winter–spring rice season.

Figure 4-6: State organisation of irrigation management in Can Tho

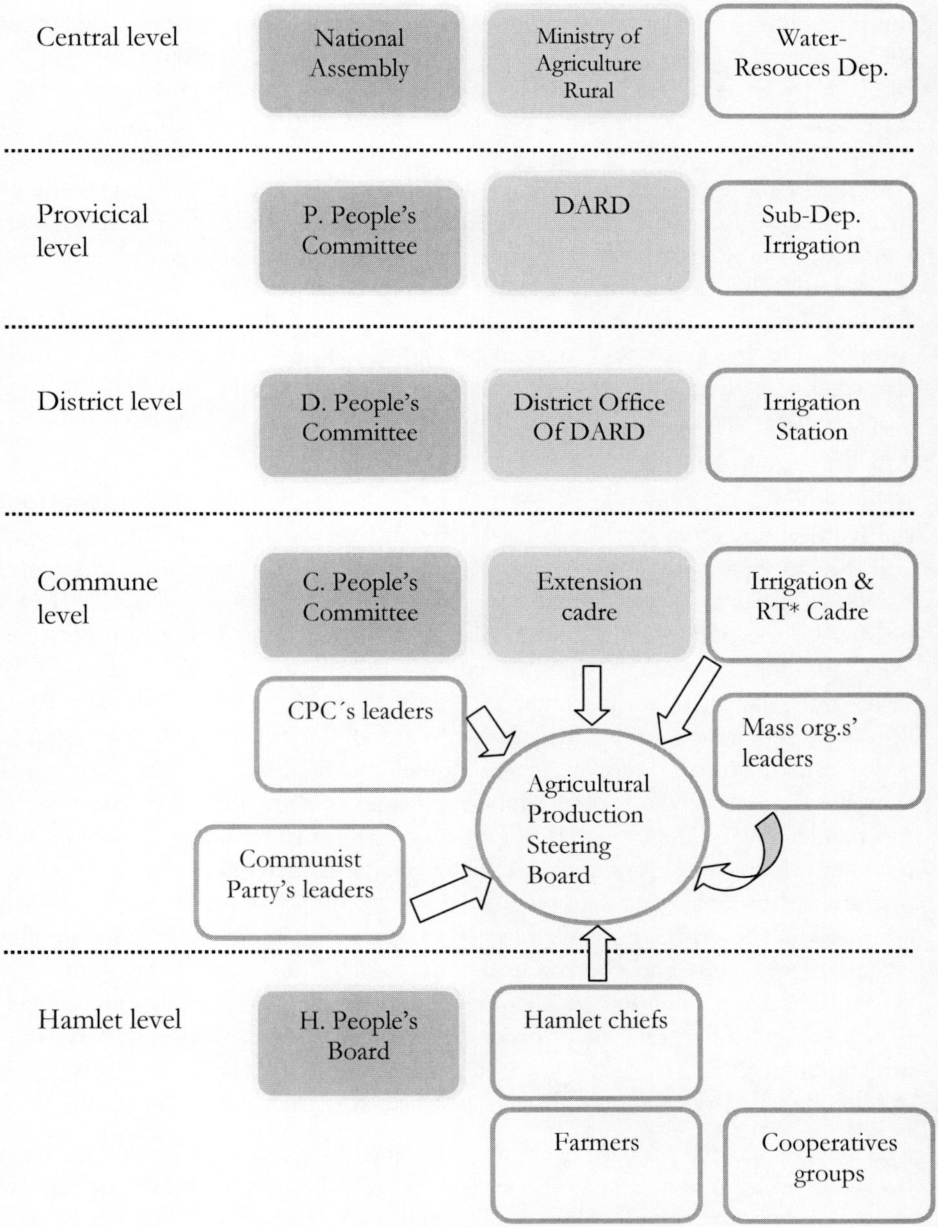

Made by author, adapted from Benedikter and Waibel (2013:16); *RT: Rural transport

Local collective leadership tends to take different forms than bureaucratic design. Although collective decision-making does occur, it does not happen all the time, and not all required cadres join the process. The question is whether these variances create an ineffective working system. According to Ho Chi Minh, '*If leadership is not based on collective action, it will cause the problems of excuses, authoritarianism, and subjectivity. Consequently, it will lead to failure*' (Ho Chi Minh Toan Tap, 505, cited in Nguyen The Thang 2012). Observation in the communes shows that non-collective decision making indeed leads to problems of excuses, authoritarianism and subjectivity, when only some hold the power to decide how to implement policies and handling local affairs. Nevertheless, those consequences might not result in management failure. In many cases, partly collective, local leadership complements other local problems of task division and inefficiency. When some cadres do not actively perform their responsibilities (see Chapter 5 on the incentives and obstacles to local cadres' work), others have to take over and skip the process of collective decision making.

Also joining the Production Board, hamlet leaders are the representatives from the hamlet bureaucracy who discuss plans and issues with the commune state and directly take care of production in the hamlet. The state structure in the hamlet consists of several cadres organised in the hamlet's People's Board headed by the hamlet chief and a vice chief. The hamlet is not considered a legal state unit, so hamlet cadres are only state assistants and receive allowances, instead of salaries. In addition, the secretariat and deputy of the hamlet's Party unit, representatives of mass organisations, and police officers make up a party–state body of up to 11 cadres in each hamlet. At this level, the Party unit plays the leading role in most activities, and the leader of the People's Board is often the hamlet chief. The Board is responsible for implementing the commune's mandates, being in close contact with farmers and their organisations (i.e. co-operatives, groups) and supporting and listening to their requests and comments. The co-operatives and groups' status, position and relation to the state, which have been debated, are further analysed in section Farmer groups as co-operative groups in the Mekong Delta. Similarly to the commune level, the ideology of 'collective leadership, individual responsibility' is also implemented in the hamlet. During the winter–spring rice production period, the People's Board appoints cadres responsible for zones within the hamlet.

The hamlet People's Board is a state-designed body and receives mandates from the commune state. At the same time, the board's members of hamlet cadres are grassroots individuals and frequently understand and partly represent the people's interests. These cadres stand at an interface between the state's policies and local interests, where their decision-making is, on one hand, very much responsive to the local conditions and interests as insiders of the farming community and, on the other hand, regulated by the state's bureaucratic format and mandates. Isolated from the practice of diversity which the superior state increasingly recognises, the hamlet's cadres frequently enjoy a certain level of freedom to make amendments to negotiate the various factors influencing a single event. The low compensatory allowance for the 'part-time' job, held alongside individual farming tasks, makes the efficiency of hamlet's cadres depend very much on individual incentives to help the state achieve its targets.

In addition, the research in Can Tho finds that hamlet cadres are both administrative and socially prestigious leaders. This status is the result of lengthy and successful state penetration into the local affairs, or, in other words, the impact of the prolonged process of interaction and adaptation between statecraft and local arrangements in the Mekong Delta.

Having focused on the local levels, it should be noted that the commune and hamlet levels are practically not part of official decentralisation. The Joint Donor Report (2010:v, 27) contends that decentralisation in Vietnam has transferred significant power to provinces but less to district and commune governments. Adding to this argument, Fritzen (2006:6) asserts that the decentralised system in Vietnam is *'"centralized" at the provincial level',* and

> the commune level — despite being the "foundation of our public administration" in Ho Chi Minh's phrase — is unlikely to see major changes in its resources or authority, given the current administrative culture of control that marks most officials at the provincial and district levels of government'. (ibid:25)

This trends explains the district and commune state's limited access to financial and other resources (e.g. dredging projects), resulting in limited authority and autonomy in decision making.

However, the limitations of the centralised system at the provincial level are compensated and, in some cases, challenged by existing local arrangements. Although organised strictly in the same format with task division, the commune and hamlet levels of state management '*do not operate in a uniformed way and the performance of state bureaucracies varies accordingly*' compared to district and provincial bureaucracy (Waibel 2010:14-15). The grassroots level is subjected to deviation guided by the local arrangements of responsibilities, which very much depend on working capacity and trust (further elaborated in chapter 5).

Management of various activities in the irrigation sector

While Figure 4-6 illustrates the basic management structure of the irrigation sector, the function and role of each agency differ according to its specific activities. This section provides an overview on the different task divisions and responsibilities of state individuals and agencies in various irrigation activities. First, this section introduces Vietnam's irrigation sector, then the irrigation of the Mekong Delta in particular and the specific arrangements in state structure to manage various activities.

Irrigation varies externally and internally among countries, regions and areas. Such variations include irrigation type, system size and the individual or group organisation of irrigation. The choice and existence of irrigation types depend heavily on local natural conditions, infrastructure availability and the social–political context of human interventions (see Chapter 3 for the history of hydraulic interventions in the Mekong Delta which has been shaped by natural, physical, social and political circumstances). The specific characteristics of the hydraulic infrastructure, cropping pattern and irrigation method co-evolve with the country's management structure, which defines the management pattern for the irrigation of regions.

Globally, an irrigation system can be run by the state or farmer or be under joint management[43]. Wittfogel (1957, cited in Fuhrmann 2008:9) describes these types of irrigation systems as existing only in situations of water shortage when irrigation can be facilitated and managed only with hydraulic works that create hydraulic agriculture — namely the communal and government regulation of irrigation. Wittfogel (ibid.) also identifies a type of water regulation system which exists only in the condition of water abundance and needs only simple infrastructure and little co-operation between single farmers. This global irrigation sector has always been context defined, challenging human efforts to categorise management.

The particular features of the Mekong Delta affect how irrigation has been set up there. In this delta, the water supplying for regional agricultural irrigation comes mostly from surface water, primarily the Mekong River and its tributaries. River runoff irrigation has been established in the areas with low elevation and a dense system of canals to deliver water to the fields through small ditch, gates, culverts and pumps. Pumps are used especially in areas with high elevation and during the dry season when water in the canals is too low to allow a continuous flow. The Delta has only a limited number of modern irrigation systems designed according to the standard definition of '*consist*[ing] *of a (main) intake structure or (main) pumping station, a conveyance system, a distribution*

[43] This categorisation of irrigation systems has been used by irrigation scholars, mostly in researches on traditional aspects of irrigation system in, for example, Thailand (Shivatoki 2000) and Nepal (Pradhan 2000).

system, a field application system, and a drainage system' (Brouwer et al. 1985). According to Vo Khac Tri (2012:76), in 2010, the Delta had 9 large-scale irrigation systems with large hydraulics works, which collectively irrigated 720,000 ha., approximately 30% of the Delta's agricultural land (roughly 2.7 million ha., 2011 data from the Rural Agricultural and Fishery Census, GSO 2011). However, operation of the projects has rarely met the expected capacity because many irrigation projects are still under construction or need to be rehabilitated (e.g. OMon–Xa No project). Consequently, the Delta is characterised mostly by surface irrigation supported by an open network of small and large earthen canals (Picture 4-1). Whether through an irrigation system, irrigation in the Delta is under state management structured by a decentralised or deconcentrated administrative system. Consequently, the scale of infrastructure does not affect the state's intent to use irrigation to intervene, access and control the local production.

Picture 4-1: Earthen canals in the Delta: Top: secondary canal or river (Huynh Linh, 17.10.2011); Bottom: tertiary canal (Huynh Linh, 20.11.2011)

Practically speaking, what is involved in the irrigation and irrigation management in the Mekong Delta? In general, an irrigation system consists of 2 main components: hardware, or infrastructure including intake structures, canals and other structures, and software, or an institutional setting to manage the system. Since the beginning of hydraulic irrigation, irrigation system is subjected to 2 main activities: operation and maintenance. In the 1990s, modernisation became active and defined by the Food and Agriculture Organisation (FAO) as

> a process of technical and managerial upgrading (as opposed to mere rehabilitation) of irrigation scheme with the objective to improve resource utilization (labor, water, economics, environmental) and water delivery service to farms. (1997, cited in Facon 2007:2)

The ideology of modernisation has manifested as the introduction of irrigation management transfer (IMT). IMT began in Mexico in the 1980s as an effort to '*reduce public expenditure whilst increasing farmer participation in the management of the irrigation systems*' (Burton et al. 2002:7). This model was then transferred to other countries. Closely linked to IMT are the formation and expansion of water user associations (WUA) and the focus on farmers' participation in irrigation management (PIM), which entered discussion at the same time with WUAs. Since then, management has become a main focal activity of irrigation, along with infrastructure construction, operation and maintenance. In the Vietnamese context, the main tasks are the management of operation and maintenance. Particularly in the flood-prone areas of the Mekong Delta, these activities facilitate drainage, which is vital not only for irrigation but also for farmers' livelihoods and safety in general and, thus, in Vietnamese are commonly called *tưới tiêu*, or irrigation drainage (Irrigation expert, interview 15.09.2011).

Irrigation and drainage in the present-day Delta involves several activities at the intersection of governance and society. First are constructing, operating and maintaining structures, such as gates, culverts and large pumping stations. Construction of large hydraulic structures, in particular, is carried out mostly by agencies at the central and/or provincial levels. At the provincial level, Sub-department of the irrigation management is representative of the provincial level. The central state undertakes the entire process for large projects, from tender organisation to financial arrangements and quality control. When the structures are in place, they are operated and maintained by paid labours who can either be under the general management office for a specific irrigation system the local authority — the Commune People's Committee, for whom the irrigation and rural transportation cadre is the main worker. The central level's monopoly of large projects is part of the established structure to maintain the power and interests of the hydrocracies (hydraulic bureaucracies) established throughout history, especially in the nation-building after the unification (see Chapter 3).

The second main activity is canal maintenance, mostly dredging of the earthen canals to restore functioning depth. The canals in the Mekong Delta are the water supply source not only for domestic usages and irrigation of rice fields and other plantations but also for other activities such as transportation and recreation[44]. Amongst those, water supply and transportation require canals of a certain depth and width. The water in the Mekong River system has a long history of being rich with substances, causing many difficulties for rulers trying to control the territory. The richness of the sediment in the canals, on one hand, provides the beneficial nutrients for crops but, on the other hand, creates loads of sedimentation in sections of the canal network which, combined with canal bank erosion, results in shallow, narrow canals over time.

Canal-dredging projects are often carried out by state-owned or private constructing firms organised and managed by state agencies at various levels. Responsibility for projects and budget management are organised according to the decentralised state structure. For the dredging budget, canals are classified into 2 groups: (1) source canals (*kênh tạo nguồn*): large canals whose maintenance is a state responsibility; and (2) field canals (*kênh nội đồng*), which farmers must dredge. In addition, Ordinance 32/2001/PL-UBTVQH10 on the exploitation and protection of irrigation works identifies 3 types of source canals: first-, second- and third-level canals. Depending on the type, a canal is assigned to a certain state level, which receives funds for dredging from specific state offices.

The ministerial (central) level is the investor in projects to dredge inter-provincial primary canals. In the dredging context, the term 'investor' (*chủ đầu tư*) refers to the project management board that provides the budget, organises and decides the technical survey before dredging, performs the tender process and quality control and approves the completion of a project. One level lower, the provincial Sub-department of Irrigation Management manages canal dredging projects restricted to the boundaries of 1 province. Additionally, in

44 Although rural inhabitants no longer commonly bathe in rivers due to (scare of) high concentrations of waste, pesticides and other pollutants, recreation remains part of the values of the waterways. Research on the loss of traditional values due to changing conditions (human intervention) can be found in Ehlert (2011).

some provinces, irrigation management companies (IMC)[45] have responsibility for source-canal projects.

Research in Can Tho shows that the district level also manages dredging projects, mostly those with costs of less than 500 million VND[46] (district irrigation cadre, interview 04.07.2011). In this case, the investor is the DARD district office, assisted by district's irrigation station. Those agencies control and organise the budget received from the provincial or district level. For example, the estimated cost of a dredging project for a 1,000-m-long and 7-m–wide canal is approximately 280 million VND (irrigation cadre, interview 18.10.2011). A canal a little longer or wider could rise to another level of management. Given certain benefits of responsibility for dredging projects, some districts tend to cut large projects into several smaller projects over which they retain management.

The division of investor and funding responsibilities for canal dredging in Can Tho are summarised in table 4-5.

---

[45] IMCs, or irrigation development management companies (IDMCs) are state-owned companies and serve as water suppliers for farmers. Formed as public organisations, IDMCs are not expected to turn a profit. They are established to service primary hydraulic units which do not necessarily correspond with administrative units. In that case, one IDMC can report to several DARDs. At the district level, district agriculture offices function as a mediator between pumping stations and IDMCs. The pumping station owners are positioned between the water users and the IDMCs and are responsible for maintenance of secondary and tertiary canals, pumping of water to fields and the collection of irrigation fees (Fontenelle 2001: 17; Hansen/ Do Hong Phan 2005: 235; AusAid 2003:19f). Private pumping stations do not exist everywhere. An alternative is a mobile pumping unit owned by a farmer who offers services to other farmers or by a group of farmers, for example, in a new cooperative (Le Meur et al. 2005: 55f; Miller 2003: 246).

[46] ~ 25,000 USD (1 USD ~ 20,000 VND)

Table 4-5: Division of management and budget sources for canal dredging

Canal type	Investor	Source of budget
Main canals and primary canals Other canals for which dredging cost more than 500 million VND	Sub-department of Irrigation Management	National budget for irrigation works – hold and distributed by the Province People's Committee
Canal with dredging costs of up to 500 million VND	DARD office, assisted by irrigation station	National budget for irrigation works distributed to the district District's budget for irrigation work.

Data from 2011–2012 field research in Can Tho.

In general, state agencies at lower levels have the task and the right to assist project investors in organising land and canal clearance and performing quality control. For instance, district offices might participate in ministerial projects, and CPCs in district projects. However, these lower-level state agencies' role is rather limited and sometimes considered merely symbolic (see section Budget management for source canal dredging for a detailed analysis). Furthermore, the designed organisation of the decentralised state structure is not necessary implemented in the same way in diverse localities. For instance in the research site, the irrigation station's role stops at managing the operation of any irrigation structures in the district, surveying the condition of canals and recording the water supply and sanitation. Although the irrigation station supposedly assists in dredging projects at the district level, the cadres have very little or no involvement in tender selection and quality control. Thus, designed management structures are often subjected to changes decided by local arrangements for task division and likely benefit sharing.

Officially, dredging field canals relies on farmers to organise and contribute to paying the costs. However, mobilising farmers' contributions to dredging projects has never been easy, especially when the costs rise due to canal widening. Chapter 5 further illustrates the on-going challenges in field canal dredging, local strategies, and the tendency to transfer all canals to the state budget, adding to the pressure of the current budget shortage.

Despite the various categorisations of canals for regulatory purposes, there is no clear-cut boundary between the responsibilities of the state and the people.

Indeed, the recent introduction of irrigation fees demonstrates how the state has attempted to balance the socialisation policy which encourages cost sharing among the people with maintenance of state control and resources for national food security and rural development goals which support the state's image of *caring for the people*. Further elaboration of this case is provided in the section on the budget for canal dredging.

The last activity closely associated with irrigation is water drainage. Amongst various activities, 1 of the 2 main targets of this research is the local field drainage before the main rice season (winter–spring season), in which practice results from a complicated process of interaction and negotiation amongst different stakeholders, including state agencies. This activity is organised and managed alongside production planning as field drainage is important for starting the most productive season. Consequently, it falls under the management of the line agencies for the agriculture sector, including DARD and its sub-departments (Plan Protection, Rural Development, Centre for Extension) at the provincial level; the offices of agriculture and extension at the district level; and the Production Steering Board at the commune level.

Budget management for source canal dredging: An illustration of decentralisation in the Mekong Delta

A large budget for irrigation in the Mekong Delta is dedicated for canal maintenance, particularly canal dredging, for which exists high, regular demand. Especially when mechanes are involved, dredging activity becomes a business venture[47]. Subsequently, the funding or state budget becomes the decisive factor in the efficiency of dredging. At the provincial level, all budgets are under the management of Department of Finance, and the provincial People's Committee — the decision makers - decide whether to fund activities after considering local requests. The whole process for requesting funds and decision-making on budget for dredging projects is illustrated in Box 1 and Box 2 below. At the beginning of each year, the provincial People's Committee decides the amount of budget to be allotted for canal dredging. The approved amount is decided based on available funds and requests received at the end of the previous year from the Sub-department of Irrigation Management

[47] Canal dredging has become a busy business not only in the Mekong Delta but also in the whole of Vietnam. Since the late 1990s, mechanised irrigation (dredging companies) has been eliminated from the monopolised system with only state- and military-owned construction companies. In addition to such companies, the contemporary hydraulics industry in the Mekong Delta includes private construction companies, cooperatives (Evers and Benedikter 2009a:17) and small-scale individual enterprises.

(interview 18.10.2011). This section describes the procedure of budget request and distribution for dredging projects at the district level under the decentralised state management structure (see box 1, box 2). An empirical account of Can Tho contributes to the analysis of the inconsistencies in budget availability, which forces local flexibility in handling dredging projects. This case also highlights the tendency towards centralisation in budget management, giving communes no control over state-managed projects.

Box 1: Steps in formal procedure for a dredging project at district level (by authors, based on interview 14.12.2011)

- DARD office receives request from communes;
- Hires consultation agent;
- Surveys all requested canals;
- Estimates the cost;
- Sends request with estimated cost to the province level;
- Decides which canals to dredge depending on the funds available

Box 2: Steps in decision-making procedure at provincial level and budget flow for a dredging project (authors, based on interview 14.12.2011)

- Office of Agriculture sends documents to provincial DARD;
- DARD transfers documents to Department of Finance;
- Department of Finance transfers documents to Deparment of Planning and Investent;
- Department of Planning and Investment submits documents to the Province People´s Committee (PPC);
- Department of Planning and Investment submits documents to the Province People´s Committee (PPC);
- PPC aproves and signs documents;
- PPC distributes budget distributed to the District People´s Committee;
- District Department of Finance holds budget.

Interviewees at the district level point to 2 regular budgets that support for the dredging of source canals. First is the regular budget for canal maintenance from the provincial state: *the irrigation work budget*. This budget is set at the beginning of the year. Second, the regular *irrigation budget of District People's Councils* is drawn from district budget, with a maximum of 300 million VND per year. In addition, other budgets added during the year. Common examples of these include the following:

- Preferential credit loans (vốn tín dụng ưu đãi): The provincial state distributes these loans to various fields, such as transportation and irrigation. Often, the available load funds are made known by May each year.
- Budget for irrigation work, including the development of new infrastructure (đầu tư xây dựng cơ bản)
- Compensating provision for the irrigation service fee (ISF[48]) (cấp bù thủy lợi phí)

Additionally, other funds are provided for canal maintenance in response to emergency situations or certain special policies. For instance, in 2011, the researched district received 2 other budgets: *a canal dredging fund for a closed dyke system* provided from the province and *a damage support fund* from the central government to reinforce the deteriorating dykes damaged by floods in 2010 (together these budgets were expected to bring 15 billion VND to Can Tho). In consideration of on-going projects, such as standard fields, new rural areas and dyke zones (*khu đê bao*), more funds are made available for canal dredging, especially in the pilot or focal communes of those projects. The irregularity of budget actually undermines the formality of arrangements for dredging by encouraging flexibility in or deviation from structured procedure and practice of setting dredging budgets.

Despite the number of budget sources, their availability is not stable, and accessibility is not certain. First, most budgets are allocated to new infrastructure, often leaving funds for operation and maintenance shortage (Bucx et al. 2010:55). Second, not all of the available budgets are known from the beginning of the year. Instead, district's cadres are informed throughout the

48 Irrigation service fees (ISFs) are defined as '*the water service charge collected from organizations and individuals using water or providing services from irrigation works for agriculture production in order to contribute to covering part of expense for management, maintenance and protection of irrigation works (Article 2 of the Ordinance on Exploitation and Protection of Irrigation Works)*' (Nguyen Thi Phuong Loan 2010a:222). The fees had been collected in Vietnam since the 1980s, but collection for water use and infrastructure in agriculture, aquaculture and salt harvesting was exempted in 2008 under decree 115/2008/NĐ-CP. Since then, the state's budget for irrigation exploitation and protection in Vietnam has included funds called 'compensating provision for ISFs'. The impacts of the policy changes have varied greatly among regions because of infrastructure and management differences. See Huynh Thi Phuong Linh and Gerke (forthcoming ZEF working paper) for a detailed analysis of ISFs policy, with empirical data on irrigation financing in the Mekong Delta.

year when more funds become available for canal dredging. Third, cadres from the 4 districts in Can Tho confirm that the present available budgets for canal dredging do not meet local demand. In the researched district, a dredging plan of approximately VND 5–7 billion for the following year is sent to DARD around December every year. However, the amount received is always smaller and varies depending on different factors, including the availability of funds and the informal dynamic of fund arrangements in the city (district cadre, interview 18.10.2011). An interviewee asserts that the supplied budget often meets only 30% of annual needs (district cadre, interview 14.04.2012). The extent to which informal dynamics effect budget allotments is unclear. This uncertainty necessitates local arrangements to handle the deficit.

The following quote illustrates the canal dredging situation at the district level under the current budget limit:

> It is a fact that canals and rivers in our delta with this sedimentation condition are often silted in a short period of time, in average from 5 to 7 years. That is when we have to dredge the canals. According to decentralisation, the first-level canals are in the city, second and third-level canals are in the district, and field level canals are taken care of by the people. Normally, we make a plan and send it to the upper level (province). We will do the work according to budget availability. If the budget is not enough, we have to wait for the following year, and if it still not enough in the coming year, we wait for the next, and so on. In case of canal which needs to be done urgently, we will mobilise the people [farmers] to do it [contribute money to dredge the canal]. (district agriculture cadre, interview 29.03.2012)

Given the budget limitations, how decisions regarding dredging projects are made at the district level is clarified by observations at the district and communes researched. In general, the district-level project management board — the Office of Agriculture — has full authority to make decisions about dredging projects within the district's boundary. Informants narrate that dredging planning and decisions are made based on several criteria (interview 18.10.2011). Most important is whether the canal is in urgent need of dredging, for example, if it is too shallow to serve irrigation and local transportation. Also, participation in in a pilot project that permits requesting funds from special programmes can place a commune's canal projects on the listed of planned dredging. In the case of many urgent canals, which happens frequently, the method of *cuốn chiếu — rolling the mattress* (made from Cyperaceae plants) — is applied. The canals are listed according to their urgency and are dredged in order, depending on budget availability. Officially, district officers have to maintain equality among communes, which means that, if one commune has canal(s) dredged this year, the opportunity will be given to other communes the next year (district cadre, interview 14.12.2011). Irrigation cadres in the communes confirmed this effort but cautioned about difficulties in meeting the local demand. The decision-making process for the canal dredging budgets remains inconsistent because of the uncertain allocation mechanism and difficulty establishing criteria to evaluate the urgency of projects.

Analysing fiscal decentralisation in Vietnam, Kerkvliet (2004:14) finds that the province receives most of its annual budget from the central level and uses '*its own system*' to allocate funds to province-wide programmes and to districts, which pass along a portion to communes. Further decentralisation, especially in financial autonomy at the district and commune levels, has been debated (ibid:15). At present, though, the district receives a specific budget for canal dredging which the district People's Committee autonomously controls, CPCs have to request that the district consider projects in their commune. The uncertainty and inconsistency in budget availability and the persistence of local centralised systems have left control over the budget and decision-making power in the hands of higher levels of the hierarchy. The responsible state offices in the communes, districts and provinces may retain a small proportion of some taxes[49] collected, including ISF before 2008, but the amount is small and directed towards all the commune's affairs, not specifically irrigation. This flexibility in accounting generally reflects the extent to which informal relations and negotiation are required to gather the resources necessary to complete all sorts of tasks, including dredging.

---

[49] Decree 24/2006/NĐ-CP (the most recent document concerning all fees and charges; for each group of charges, there are more recent regulations) reported 73 kinds of fee in 12 categories and 42 kinds of charges in 5 categories. Only a small number of fees obligatory for all households are collected annually by the local government. These fees are regulated by provincial People's Committee. In Can Tho, hamlet cadres collect the Fee for Security and Defence (48,000 VND, ~ 2.20 USD/household/year), the Fee for Preventing and Coping with Disaster (storms and flooding) (10,000 VND/household/year). In addition, each household pays a housing land tax (22,000 VND/household/year). The number of fees varies by periods. For example, other fees were collected in rural areas after 1975: IST (until 2007–2008), Fund for Building Rural Areas (or Building the Homeland, until 2003) and Fund for Gratitude (for people and families who contributed to the war effort, until 2009). The impact of these fees to household income varies between small and large-scale farms. As stated in an articles on Thanh Hoa (Nguyen Thi Hien 2008:149), '*the total payment by a farm household is not really big, as compared to that of big-scale farms, better off families or an urban household, but it may make 50% or more of the total income of an ordinary peasant household*'.

In summary, although decision-making in dredging projects lies in the hands of the responsible offices at the level to which the project belongs, the incomplete fiscal decentralisation leaves a centralised structure in place at the provincial and district levels and grants control and power over budget decision-making to higher levels of the state. In general, the budget for irrigation maintenance is inadequate because of an insufficient state budget and unstable contributions from the water users (generally farmers; field canal dredging is officially categorised as farmers' responsibility). The budget uncertainty and difficulty in establishing criteria for decision-making make the mechanism for budget allotment inconsistent. This inconsistence suggests the involvement of informal social relations and allows local agencies flexibility in using the available budget, even to applying the state budget to the dredging of field canals, when possible. On one hand, these local processes might allow room for corruption or inequality between localities by favouring certain projects based on personal relations. On the other hand, these local processes can be seen as adjustments to best use the available but inconsistent and limited state budget to fulfil state mandates and local demands. This process of organising the dredging projects also highlights the existing state structure that allows local deviation, for instance, to use the state budget for field canal dredging.

Given the budget shortage, one might anticipate the deterioration of irrigation infrastructure, which could create critical obstacles to the irrigation, drainage and local transportation. That is the foreseen impact of budget shortage and low incentives for farmers' contribution as a consequence of the abolishment of ISF in Northern Vietnam (irrigation expert, interview 15.09.2011). However, the Mekong Delta, which is less dependent on modern' hydraulics infrastructure, might not be subjected to the same impacts as the Red River Delta. Furthermore, the state has a significant concern in maintaining the functionality of irrigation infrastructure (mainly canal dredging) in the Mekong Delta in order to support food production. Government Resolution 63/NQ-CP on securing the national food security (2009) states that '*the issue of food security is especially concerning for and guided by the Party and the Government in order to secure nation food security in short and long term*'. This concern might ensure continuous investment into Delta's hydraulic landscape, but it does not necessarily mean that all communes and farmers benefit fairly from it.

Discussions and conclusions on decentralisation in irrigation management

Research on decentralisation in irrigation management and budget management in the Mekong Delta illustrates the current situation and explains the rationales for the technical and political behaviour emerging from decentralisation in

Vietnam. In general, there are 2 main barriers to implementing fiscal de-concentration down to the commune level and devolution in general. First is the culture of control over agencies' resources at the higher levels of the hierarchical ladder. This culture has been characterised in the research about water bureaucracies – hydrocracy (Benedikter 2014b). Benedikter asserts that '*the strong legacy of central planning and the ideology of state-centrism facilitated the adherence to technocratic top-down implementation and a bottom-up reporting mechanism*', which, continued in the era of water reform, is an illustration of '*pouring old wine into new bottles*' (ibid:152-153). Indeed, the coordinating system of local satellites which seeks to ensure the effective implementation of centrally directed hydraulic missions is actually a form of administrative decentralisation (ibid:47). However, central bureaucrats still retain monopoly power over large irrigation projects, including those funded by international donors. In another case, the top-down style of centralised management of dyke planning continues with weak bottom-up influences (Pham Cong Huu 2011:71). The limitation of decentralisation is also illustrated in the dredging projects at the provincial and district levels (this research in Can Tho, 2011–2012). The local state has only a symbolic role in the planning and implementation of dredging projects; for instance, the DARD district offices have no role in tenders, budgeting and quality control in the projects managed by the Sub-department of Irrigation. Neither do the CPCs and farmers in district-level projects.

Second are the generalisation used to justify the limited authority and autonomy given to the lower levels of the state. It is claimed that the local level has not yet become capable of managing affairs with a high degree of autonomy. Comments such as '*not well trained*' or '*very limited*' knowledge of the local markets and law, '*poor organisational and managerial capacities*' (Kerkvliet 2004: 12) were constantly mentioned in this 2011–2012 research. The risk of empowering a form of local '*self-governance*' due to decentralisation threatens the central state, so '*local administrations need to be put under the control of the central state (again), if uniformity and consistency in the translation of national policies should be achieved*' (Joint Donor Report 2009:ii, cited in Beresford 2006:15). The Vietnamese state has never abandoned its goal to have a certain level of control over practice, either through the command-and-control strategies applied during the 1970s and 1980s, the dense, contemporary state structure facilitating mobilisation or what is called the state's application of infrastructure and discursive power, which are analytically explored at the local level in this research.

Overall, decentralisation in Vietnam, particularly in the realm of irrigation management, is constrained by persistent, internal features of power and hierarchical arrangements. However, Benedikter (2014b) holds that '*administrative pluralism and diversity in water control and infrastructure management is the immediate consequence of a clash with the ideology of a centralized state system*'. Indeed, despite state attempts to centrally control budget management, the inconsistent budget system has allowed local cadres flexibility in acquiring many resources at that level. Budget management remains strongly centralised in order to benefit the group interests of hydraulic bureaucrats.

Irrigation management as the business of more stakeholders

The state apparatus to manage irrigation and the agriculture sector cannot single-handedly realise state goals, despite its dense structure of agencies, detailed regulations and legal documents. In communes and hamlets, practices are guided, instead, by interaction amongst stakeholders, of which the party–state agencies are part. Amongst the others, the most important are farmers in groups or co-operatives who are generally the final objects of state policies and the masters of social-economic issues according to the state slogan: '*The Party leads, the government manages, the people are masters*'. Also joining and significantly influencing the local practices are the increasing business stakeholders and organisations that stand in the space between a party–state agency and a civil organisation. This section introduces these non-state or state-related actors and their historical positions in relation to the bureaucratic state structure for irrigation management. It is illustrated that, at the level of practice, the state is not the sole player, and all actors are in one or another way connected to each other. Their connections and the local arrangements and institutions which manage the sector are analysed in Chapter 5, focusing on the case of field drainage and field canal dredging.

Agencies located at the bureaucratic–informal interface

The irrigation management system includes agencies whose role is to represent and work closely with farmers, although officially they are accountable to and, to certain extent, under the management of either the state or the Party. These agencies are extension office and mass organisations. This section discusses the status of extension offices and mass organisations, which confuse both Vietnamese and international agencies as to the extent to they are state agencies or civic organisations. For instance, while mass organisations claim to be people's organisations supposedly formed by the people and operated for the people's interests, the term 'people' is a well-known part of the Communist Party's vocabulary. This connection illustrates the relation of these organisations to the party–state system.

Extension office

An official state but more people-oriented organisation is the extension centre. The Department of Extensions for Agriculture and Forestry (*Cục khuyến nông khuyến lâm)* was established in 1993. After the establishment of Ministry of Aquaculture in 2000, the department was separated into 2 centres in 2003: the Centre for Aquaculture Extension and the Centre of Agriculture Extension. Since the 2007 merger of the Ministry of Agriculture and Ministry of Aquaculture into MARD, all extension services have been placed under the National Centre for Agriculture and Aquaculture Extension. Following the de-concentration of state management, the Offices of Agriculture and Aquaculture Extension are distributed across levels, leading to provincial and district extension centres with only 6 to 7 staff members each. CPCs offices, in particular, have 1 or 2 extension cadres who receive their salary from district extension stations.

The main tasks of agriculture extension offices are to provide agricultural technical knowledge and training and to transfer technologies. According to Nguyen Duy Can, an agricultural expert at Can Tho University (personal communication 2011), the main activity of the extension offices in the Delta today is to implement technical demonstration models. The extension centre sets up demonstration sites at several communes according to programmes initiated either by the national or provincial centre, sometimes in co-operation with companies. To a certain extent, extension agents in the commune have the same status as other commune state professional cadres, for instance, the irrigation and rural transportation cadres. However, while helping farmers devise agricultural strategies, the extension agents have a strong responsibility in *building the bridge* (*làm cầu nối*) between the farmers and the district's extension station, between farmers and the commune state. It is argued by Hicks (2005:230) that the '*extension officers are positioned at a blurred interface between society and the local state*' and are the '*associates of the state*', rather than state actors. In general, although the extension officers are an official state activity and part of the ministerial apparatus, the extension officers make their role a more people-oriented activity.

Mass organisations: The Party's prolonged arm or agents of the masses

Mass organisations, on the one hand, are social–political organisations which under the scope of Communist Party and are present at every administrative level, '*ensuring the effective education of the masses*' (Reis 2013:18). Simultaneously, they represent various social groups of inhabitants, intervene to protect their rights and transfer their requests to the state. The position of mass organisations has been somewhat relatively and functioned differently during historical periods since their establishment in the 1930s.

The term 'mass organisation' refers to social–political organisations *(Tổ chức chính trị xã hội)*, which are collectively known as *Đoàn thể* in Vietnam. The country has 44 mass organisations, including the Women's Union, Ho Chi Minh Communist Youth Union, Farmer's Union, Vietnam Confederation of Workers and Veterans' Association. The Vietnam Fatherland Front has acted as the general management agency for the mass organisations, and its member organisation includes the Vietnam Communist Party. The front's logo represents the nation and all its groups united to build a Vietnam with a '*prosperous population, strong nation, society with equality, democracy and civilisation*' (Introducing the Vietnam Fatherland Front[50]). The role of mass organisations has varied among periods of socio-political and economic changes. In the 1930s, immediately after the First Congress of the Communist Party, the organisations aimed to gain the participation of various groups, such as farmers, women and youth, in order to create a strong, united power for solidarity (Nguyen Trung Vinh and Nguyen Khuong Ba, 2008:34). Later, in the on-going struggle to gain control over territory, the mass organisations changed their names and, under central supervision by the Viet Minh Front, mobilised inhabitants to contribute to social, economic and political actions. . To save the nation during the period of 1945–1946, mass organisations aimed to form grassroots government, eliminate illiteracy and reclaim and rehabilitate cultivated land to increase production and reduce starvation. The organisation also joined the war against French colonisation until 1954 and later American intervention until 1975 (ibid). In short, mass organisations have always been hybrid organisations, serving state purposes with a participatory orientation.

Since reunification in 1975, mass organisations have constantly mobilised the participation of inhabitants to implement state policies and follow the ideology and development paths of the Communist Party. For instance, the Farmer's Union's role is described as follows:

> The Farmer's Union at all levels takes the advisory role to Communist Party units and People's Committees, guiding the farmers in following the state's policies and the Party's strategies in the current revolution towards Socialism. Also, the Union understands the farmers' aspirations and capacity in order to decide suitable strategies to achieve targets and tasks and successfully implement the Party's policies in the present modernisation of agriculture. (Nguyen Trung Vinh and Nguyen Khuong Ba 2008:378).

---

[50] http://www.mattran.org.vn/Home/GioithieuMT/gtc4.htm, accessed 17.09.2013.

In the present day, mass organisations and their cadres are present at all administrative levels, from the central state to the province, district, commune and hamlet. Mass organisations are bureaucratically organised in the same fashion as the state structure with double subordination. Each mass organisation unit is responsible for reporting to the same union at the next highest level and to Party and/or state management agencies at the same level. Given the rigid bureaucratic structure and the task of disseminating state and party policies, mass organisations are '*seen as the "civil arm" of the party apparatus, through which the political elite propagates its views on virtually all aspects of life*' (Reis 2013:19). Until the early 2000s, 74% of Vietnamese citizens were members of at least 1 mass organisation and, on average, 2 to 3 organisations (Norlund et al. 2006). However, in this study, field research in the Mekong Delta shows that official membership records of mass organisations and statistical data in general do not reflect the actual situation. In addition, distinguishing between mass organisations' units and farmers' groups can be confusing, causing mistakes in the membership record of unions at the local level. As well, membership in a union does not ensure active participation. However, to certain extent, membership does increase opportunities to access technology transfer programmes (e.g. technical training, companies' pilot contracting programmes offered in co-operation with the Farmer's Union). In general, membership in mass organisations signifies their utility. By joining organisations, members can benefit from their programmes, especially training and technology transfer.

This research conducted in Can Tho, Bac Lieu and An Giang (2011–2012) shows that today, mass organisations performs various tasks. They implement programmes and policies regulated by the state and the Party and proactively assist in production and internal local affairs. Moreover, unions and the Fatherland Front play an important role in nominating candidates for National Assembly and People's Council elections. This role is in line with their presence in the People's Council and Party units at all levels. Thus, mass organisations constitute the '*third component of the Vietnamese political system … link[ing] major socio-economic sectors and interest groups in society with the party*' (Porter 1993:87).

All mass organisations claim to represent various groups of inhabitants. Unions aim to understand people's aspirations, transfer them to the state and the Party and protect the right of the inhabitants. Consequently, mass organisations have been considered to be the formal entry point of farmers, women, youth and other groups into political space and the modes of organisation for collective action in rural areas (Larsen 2011:317). However, some see this function of mass organisations as intended to allow the Vietnam Communist Party to assess the moods and attitudes of these groups (Le Duc Tho, cited in Porter 1993:87). Given these 2 roles, mass organisations in Vietnam are '*quasi-official structures*' (Porter 1993:88), or in other words, these agencies are located at the interface between the state–party and citizens. Standing at this interface, unions are intended to fulfil various requirements and demands from both the state–party and from the people. Therefore, similarly to many other agencies at the interface, mass organisations have differentiated accountability which facilitates the diversity at the local level and the rigid bureaucratic state–party structure of management.

Both extension offices and mass organisations have a dual accountability to state mandates and to inhabitants' aspirations. To this group, the local state, including CPCs and hamlets' People's Board, can be added because the local state serves as an intermediary connecting higher-level government agencies, with their policies and mandates, and the local people, with their everyday wishes. The questions are to which part of the state or society each agency feels it belongs and whether mass organisations can have a role in the civil society. This work, particularly Chapter 5, can guide the debate to analyse the space where various individuals and agencies interact and negotiate, seeking to fulfil their interests and mandates. This analysis does not try to draw a strict line between the state and the inhabitants because the boundary is blurred, or in other words, the Vietnamese system has never intended there to be such a line. Instead, the analysis focuses on the negotiable position(s) the actors take and the decision-making processes leading to local practices.

Farmers and farmer organisations

Nowadays, farmers not only are the masters on their land, reliant on farming to provide their household's livelihood but also provide food for local (and national) consumption, serving food security and export. Consequently, farmers have been objects targeted by state programmes and policies, especially those related to agricultural production. However, the tools at the disposal of the state are rather imprecise, and farmers in the Mekong Delta are a heterogeneous community, which makes it challenging to design interventions comprehensive enough to serve all groups. First, the Delta's diversity has various ethnic groups: Kinh (commonly called Vietnamese), Khmer and Chinese. There are small and large landholders, as well as landless farmers and agricultural wage labours. Farmers also differ by cropping pattern: rice farmers, orchard farmers and multi-tasking farmers (e.g. farmers, businesspersons, part-time workers and local state's assistants). Beyond these characteristics, the Delta's inhabitants and landscape possess many other social-production features resulting from the historical dynamics of social and economic change involving natural resources (from adaptation to control) and technological advancement (as detailed in Chapter 3).

In addition to an individual basis, the farming community in the Mekong Delta today remains organised in agricultural co-operatives (*Hợp tác xã*) and other types such as clubs (*câu lạc bộ*) and groups (*tổ*) with different levels of involvement. The organisations supposedly organised by the farmers themselves have become a primary avenue for state access and intervention. A former president of the Vietnam Co-operative Alliance states:

> Through the activities, co-operative groups and co-operatives have not only confirmed their important role in production and business but also become an irreplaceable factor contributing to boosting democratisation and ensuring social welfare and political stability at the grassroots level. (Nguyen Tien Quan 2008, translated by author)

This statement is suggestive of the Vietnamese state's view of the irreplaceable value of organising in co-operatives, in effect retaining the benefits of a collective economy. State officials' appreciation of farmer organisations, whether or not formally expressed by the state, was widely confirmed in interviews and visits to government offices at the provincial, district and commune levels (2011–2012 research in Can Tho).

Amongst farmers' organisation, there exist significant differences between co-operatives and other forms, such as groups and clubs, in scale, level of organisation, legal status and extent of state involvement. Co-operatives are considered enterprises with a legal status under the Law on Co-operatives 2003. Based on the history of the form, a management structure is already set up. In Can Tho, co-operatives register in order to receive technical support from the Co-operative Alliance and legal support from the Sub-department of Rural Development. These 2 agencies also handle technical and state management of co-operatives, respectively. According to state cadres, the co-operative is the advanced stage of groups and clubs (group interview 01.02.2012). As such, the state and state-related agencies have higher level of engagement with co-operatives than groups.

Although co-operatives are legally defined and regulated, the group sector continues to be unclear. What are these groups? How have they been formed, organised and managed historically in the Mekong Delta? And how are they categorised in relation to state and civil organisations?

Given this research's focus on the local management of irrigation which is strongly related to groups, this section discusses this smallest form of organisation in the contemporary Mekong Delta. The various forms of groups are analysed in order to clarify and analyse their diversity in form and name. Next discussed are the historical formation and activities of the groups which has been closely linked to local infrastructure and social–political and economic changes. The analysis reveals differences between the contemporary and the 1970s–1980s form of collective organisations and how the state and farmers have chosen to support different forms of farmer organisation or, in another words, have adapted and modified the way of working in agriculture in the contemporary era. Data were collected from ethnographic research of observations and interviews in the 3 provinces of Mekong Delta, supported by in-depth analysis of the 3 communes in Can Tho. This empirical account of groups helps describe the diversity of groups, arguing that they stand in the space between state influence and local arrangements and are autonomous from the state. Also, this analysis on groups aims to partly fill the gap identified by Waibel and Benedikter (2013:3–4) by conducting *organisational studies 'from within'* but leaves open the question of '*how these organisations operate and how they manage their interfaces with the state'* (ibid) for Chapter 5 to answer.

Farmer groups as co-operative groups in the Mekong Delta

The term 'co-operative group' ('*Tổ hợp tác*') refers to an organisation of farmers with either an informal or signed agreement/contract to work on one or more activities. As demonstrated by Vietnam scholars, inhabitants organising for certain purposes is not alien to Vietnamese villages. Nguyen Xuan Tiep (2008:26) states that a system of people's contributions or human-resources mobilisation for irrigation and drainage was already popular during the Le and Nguyen dynasties in the 18th and 19th centuries, as found by the large excavations). Today, groups consists of organisations at the commune or hamlet level with different forms, both profit and non-profit, serving economic (e.g. agricultural production), social, cultural or a combination of these purposes (Waibel and Benedikter 2013:7).

The 1997 co-operative law marks the first legal recognition of groups. Later, the Civil Code 2005 (33/2005/QH11) officially defined co-operative groups (*tổ hợp tác*) as

> formed based on a co-operating contract which is certified by the commune People's Committee. Each group has more than 3 individuals, contribute asserts and labour to work on certain activities, together get the benefits, take responsibilities and is an object in civil relation. (Ministry of Justice 2005:189 – Article 111, translated by author)

However, the state's management role over groups was clarified only in 2007 with the approval of Decree 151 on organising the operation of co-operative groups. The decree regulates group registration in communes. Since then, groups have been legalised by a decision issued by the CPCs (Figure 4-7), which supposedly requires a working contract signed by group members. The contract should be renewable and valid for a year or a season, in the case of seasonal groups such as pumping group, and the membership should be updated annually. Interviews in Bac Lieu, An Giang province, and Can Tho confirm that Decree 151 has given groups a legal status within the state bureaucratic structure (interviews 19.07.2011; 22.07.2011; 28.06.2011, respectively). However, not all groups have an official decision paper, and not all legalised groups are active. The group situation in the researched area is more diverse than the officially unified group structure.

Visits to 24 communes, including 3 in-depth visits, show that the process of legalising groups with decision papers has had little impact on groups in the Delta, and annual updating has never been fully practiced. Evidence from the pumping groups illustrates that, although groups with various land sizes and numbers of farmers have been registered at the commune according to Decree 151, the decision papers drew attention only when new papers were needed for the implementation of the new rural area in early 2012. The necessity of the paper was not mentioned during the visits to 24 communes earlier in 2011. In this particular realm, the state's attempt to render groups transparent and governable faces typical bureaucratic hurdles while people continue to organise informally. Most local actors, including the state cadres, see little difference between groups with and without legal registration if they are only formed once a year and have no other activity besides water drainage before the sowing of the winter–spring rice season. Without registration papers, farmers still co-operate with each other through facilitation by hamlet cadres or farmers themselves. The area and number of farmers involved in the co-operations varies annually, depending on the results of negotiating the sowing period and bargaining behaviour for the co-operation of farming individuals.

Figure 4-7: Example of a decision on group establishment issued by the Commune People's Committee (translated by author)

PEOPLE´S COMMITTEE DONG THUAN COMMUNE	PEOPLE REPUBLIC OF VIETNAM Independent-Freedom-Happiness ---------------------

Dong Thuan, 21 November 2011

DECISION

Regarding the establishment of economic cooperative group number 3 – Dong Hien hamlet

PEOPLE´S COMMITTEE DONG THUAN COMMUNE

Based on the People´s Council and People´s Committee Law date 26 November 2003;

Based on Decision 4963/QD-UBND date 28/06/2011 of the Chairmain of Thoi Lai district People´s Committee: Regarding the approval of election result for the member of People´s Committee Dong Thuan X, period 2011-2016;

Considering the suggest of Dong Hien hamlet´s head;

DECIDE:

Article 1. Now, establish the economic cooperative group number 3 of Dong Hien hamlet year 2011 including these members:

* Group head: Mr. Tran Van Uon
* Vice head: Mr. Pham Van Chinh
* Secretary: Mr. Phan Tuan Dat

Article 2. Cooperative group Dong Hien has the duty to mobilise the households in their group to carry out certain activities such as: collective field drainage, sowing at the same time, support each other in capital and technologies, and product sale.

Article 3. The office of People´s Committee, and the people named in article 1 has the responsibility to implement this decision from the date signed

On behalf of the People´s Committee
Sign for- Chairman – Deputy Chairman
Signed and Sealed
Nguyen Huu Dat

Although groups supposedly are regulated by several legal documents and managed by various state agencies and mass organisations, confusion exists amongst the about who has what responsibility. Consequently, groups continue to operate according to the local governance structure, which varies among localities and is influenced by state management and other local arrangements. Data collected from An Giang, Bac Lieu and Can Tho illustrate that mass organisations at all levels, mainly the Farmer's Union and the Women's Union, are the main actors in mobilising groups' establishment and supporting groups in operation and in conflict and crisis situations (Figure 4-8). For state management, party–state policies and programmes name the Sub-department of Rural Development (*Chi cục phát triển nông thôn*)[51] at the provincial level, district DARD office and the CPSs as the management agencies overseeing the organisation of groups. In addition, the Co-operative Alliance, as the main manager of co-operatives at the provincial level, gives technical support and facilitates and guides the upgrade from group to co-operative.

[51] This agency's old name is the Department of Cooperatives and Rural Development.

Figure 4-8: State structure for managing co-operative groups in Can Tho (designed by author based on 2011–2012 interviews)

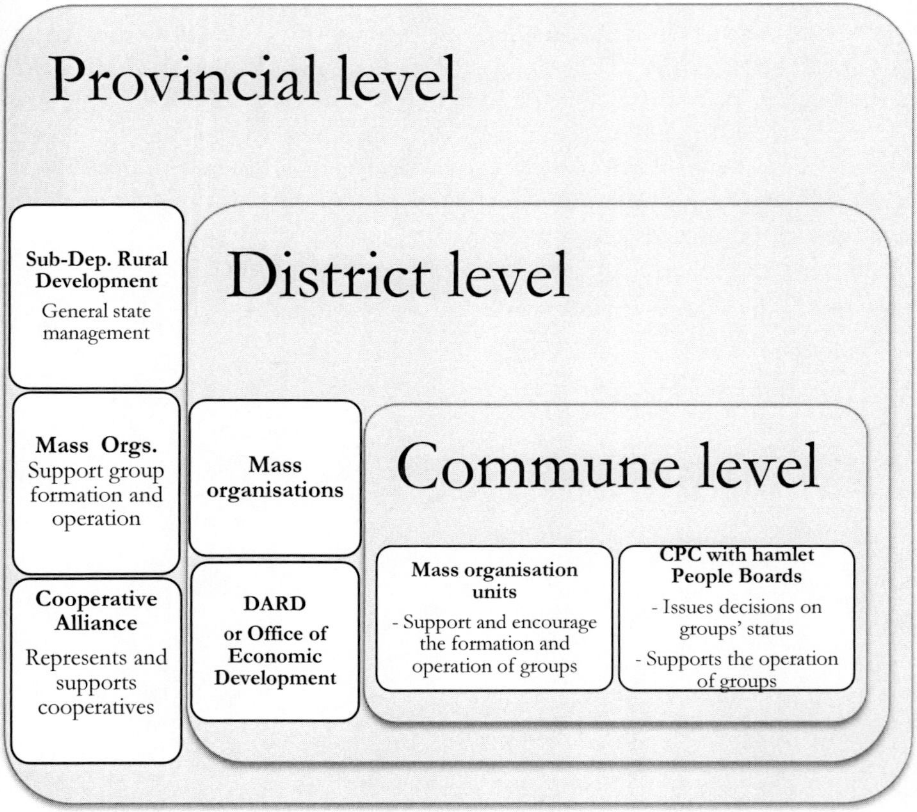

Based on this research, a clear set of regulation or task division for managing groups in Vietnam is not in place. At the beginning of the field research in 2011, I had a very blurred idea of who was doing what among the various agencies listed in Figure 4-8 for the 3 provinces and city. The conclusion from the meetings and interviews in the later period steams from the lack of a clear task division among agencies for who and which offices take which responsibilities. This confusion derives from the non-uniformity of groups, their impacts on farming practice and the unclear state system for defining groups, explained in the following sections.

Diversity and confusion in groups and group management

Research in the communes in Can Tho illustrates that, despite of being an object of state management, the management of groups also remains an object of local arrangements; in other words, the group is still an informal matter, more or less outside the state bureaucratic structure. The first issue is the uncertain record of the number of groups. State cadres at the 3 administrative levels (provincial, district and commune) claim that the reported numbers for groups — 695 groups in Bac Lieu (interview 22.07.2011), 709 in An Giang (interview 19.07.2011) and 2675 in Can Tho[52] — do not reflect the reality on the ground. The reason is clear. Many registered groups are pumping groups or others which are seasonal; consequently, their area and membership change every year depending on the water level and willingness of the farmers. However, the CPC's decisions about groups are not renewed or updated every year, making the recorded data out of date. This situation was confirmed by the local cadres in all 24 researched communes in Can Tho. A provincial cadre in An Giang explains that information on the number of groups cannot be provide as '*having a complete data on how many co-operative groups is impossible*' (interview 17.09.2011); officials gave similar denials for information in Bac Lieu and Can Tho. As a result, the statistics on the co-operative groups — for instance, that there are approximately 300,000 nationally, of which more than half work in the agriculture sector, as stated by Nguyen Van Nghiem from the Department [*Cục*] of the Co-operative Economy and Rural Development under MARD[53] — are only estimations.

The uncertainty in groups' records does not result from weakness in state management. Rather, the cause it the nature of the groups which have a high level of diversity and stand at the space between state-managed and self-managed organisations. At the surface level, the diversity of groups is shown in the name 'group'. The term 'co-operative group' mentioned in the legal documents is neither the unique nor the common name for groups in Vietnam. For instance, groups are listed as water user groups[54] in donor projects (Nguyen

---

[52] Data are taken from the 2010 report on the progress implementing the development plan for the collective economy of Can Tho in 2006–2010. The Cooperative Alliance lists another number in all reports from 2008 to 2011 – 3371 groups. Overall, it is unknown exactly how many groups there are.

[53] According to Nguyen Long (2012, in agroviet.gov.vn).

[54] A group of scholars, including many from the Centre for Participatory Irrigation Management (CPM) formed in 2004 in Ha Noi, has studied water user groups in Central and Northern Vietnam. Many groups have been established, along with the projects funded by foreign donors such as Asia Development Bank (ADB) and Agence Française de Dévelopement (AFD). However, the PIM model for groups tends to return to the starting point after the completion of the projects, regardless of their scale and funding sources (Nguyen Xuan Tiep 2008:81).

Xuan Tiep 2008b, Benedikter and Waibel 2013), while Norlund et al. (2006) describes community-based organisations as part of Vietnamese civil society[55]. Groups with the same name may differ in activities, and frequently, various terms are used for the same type of group. The diverse nomenclature for groups results from the same process of interactions between built-in local elements and the discourses of party–state interventions. These varied terminologies are as familiar as the co-evolution between development discourse and existing arrangement of Vietnamese party–state system and are a piece of traditional village and regional governance in the civil society vernacular in Vietnam (Le Trong Phuong 2013:172). To show the diversity of groups and their names, a summary of names addressing and related to groups in Vietnam is given in Table 4-6.

55 It is not within the scope of this book to discuss whether the various forms of groups/associations in Vietnam are part of civil society or a form of water-user groups. For more on this topic, see Waibel and Benedikter (2013a, 2013b).

Table 4-6: List of recorded groups in recent researches on groups in Vietnam

Legal name	Recorded Names	Source
Co-operative groups	Water-user groups Extension groups Saving groups Credit groups based on formal or informal credit (Helvetas 1996) Informal groups for agriculture, neighbourhoods, dancing, sports, festivals and celebrations Cultural groups Senior adult groups	Norlund et al. (2006) — assessing civil society in Vietnam
	Water-user associations and groups Irrigation groups Irrigation teams (under the umbrella of agro-co-operatives) Waterway teams (in the An Giang province) Mutual–help associations — in building houses, ploughing and transplanting (named in Vietnamese *Phường*, *Hội* and *Yểng*) Labour exchange groups	Nguyen Xuan Tiep (2008:28, 81–82) — researching various forms of PIM with surveys in 29 Vietnamese provinces
	Production groups (specialising in rice, gardening, fishery or aquaculture) Irrigation and pumping groups Micro-credit groups	Benedikter and Waibel (2013) — surveying groups in 8 communes in Can Tho in 2009–2010

My 2011–2012 research in the Mekong Delta faced the same obstacle when informants gave various names but little explanation of the distinctions between them (see Table 4-7 for the list of group names recorded in the 3 provinces). Most groups are formed locally, initiated either by farmers themselves or the local state. The same type of groups can be named differently by farmers, state cadres, state agencies and localities. However, influence from the policy-level state and other actors, including local cadres, prompts change in group names.

Table 4-7: Group names recorded in Can Tho and Bac Lieu and An Giang provinces, 2011–2012

Group name	Used By	Location	Actual Situation
Pumping groups (*Tổ bơm tưới*)	Commune, hamlet cadres and farmers	All researched provinces	Conduct field drainage before the main winter–spring rice season Formed when water starts receding and dissolved when pumping-out is completed
Production groups (*Tổ hợp tác sản xuất*)	State cadres	All researched provinces	All kinds of groups co-operating in production processes, such as rice production, orchard production, salt production, production of safe vegetables (*rau an toàn*) and the construction of the closed dyke system (for rice production in Bac Lieu), plant breeding and pilot production models
Waterway groups (*Tổ đường nước*)	Hydraulics management and related offices	An Giang	Pumping groups

Production groups (*Tổ liên kết sản xuất*)	Farmers and Farmer's Union officials	An Giang	Traditional and locally developed name co-operative groups[56]
Co-operative and mutual–help groups (*Tổ hợp tác tương trợ*)	Farmer's Union (listed in group records)	Can Tho	Pumping groups with no mutual–help activities
Mutual–help groups (*Tổ tương trợ or Tổ hùn vốn*)	Hamlet cadres	All researched provinces	Groups in which each member to a common fund used to give loans to the poor members to invest in production activities; voluntary contributions more idea; than real practice[57]
Loan groups	Commune and hamlet	All researched	Groups which benefit from the loan programmes[58] and/or

[56] An experienced official from the provincial Farmer's Association of An Giang asserts the groups' development changed in the province over time, as follows (interview 20.07.2011). From the collectivisation period until 1990–1992, the name *water way group* was applied to groups cooperating in the water pumping. From 1992 to 1997, local (provincial) state mobilisation (under provincial Directive 25 year 1991) encouraged the formation of *production groups* (tổ liên kết sản xuất), resulting in a massive number of such groups in An Giang, peaking at 8,000 in 1995. This period is regarded as the golden time for these groups. The state recognised the model as the rationale for the legalisation of groups founded under the 1997 cooperatives law. In response to the nationalisation of groups' names, favour was given to a new national name for group. Thus, the An Giang *production groups* were officially renamed as cooperative groups in 2007 (under the provincial directive 01).

[57]A former leader of a commune Farmer's Union in Can Tho (interview 30.12.2011) and a mutual-help group states that collect the money from members was not easy, and he had to apply certain strategies not revealed here in respect of the interviewee's request for confidentiality of the information. The model did not last long due to the loss of funds over time.

[58] The loan programmes serve to alleviate poverty. In the programmes, banks, mainly the policy and social bank and the agriculture bank, give loans for poor households to

(*Tổ vay vốn*)	cadres	provinces	trainings under the management of mass organisations, especially the Farmer's Union and the Women's Union Reported to have no collective action between members and to be deactivated immediately after the loan process
Agriculture group, self-management and Security groups (*Tổ nông nghiệp, Tổ tự quản hay tổ an ninh tự quản*)	Hamlet and commune cadres	Can Tho	Pumping groups Self-management or security groups which manage production (including pumping) activities and security in their hamlet (3–4 groups per hamlet) by checking residents in and out and reporting security-related issues to local authorities (mainly 1980s–1990s)

* Note: This research focuses on irrigation activities; therefore, most of the groups listed in this table are affiliated with the Farmer's Union. Other groups and group names exist in the researched areas, such as extension clubs and micro-credit groups.

Further exploration in 24 communes in Can Tho confirms that groups' names do not necessarily portray their nature or activities. Frequently, members of production groups only work together in field drainage before the winter–spring rice season. The mutual–help aspect of co-operative (and mutual–help) groups today only appears on paper. For instance, a number of groups were established only for the purpose of getting loans from the social-policy and agricultural banks (more details on the case are analysed later in the case of Xuan Thang commune presented in this section). The confusion in group names also appears in the interviews with local cadres; sometimes, the

invest in the production activities. Group leaders and the Farmer's Association or the Women's Union take on the role of representative and guardian for the loan process.

confusion is deliberate in order to report a significant number of groups in the commune as a sign of commune competitiveness. That tendency not only results in unreliable data on groups that bureaucrats collect from commune cadres but also challenges researchers using ethnographic methods. The following in-depth study of 3 communes in the Thoi Lai district of Can Tho contributes to the analysis of the nature of groups.

Nature of groups and the state's role

Regardless of whether groups are within the party–state system under the management of the state agencies or an informal, autonomous form of gathering and co-operation among inhabitants of Vietnam, this research calls for another approach which views groups as actors standing at the bureaucratic–informal interface where occurs negotiation among the party–state, related agencies, farmers and other actors.

On one hand, as propagated and believed by the socialism-oriented government of Vietnam, groups are part of the collective economy managed and regulated by various levels of state agencies and by mass organisations associated with the Party. The socialist-style support is expressed through promoting groups by offering incentives, including technical and financial supports (Kono 2001:77–8). On the other hand, groups are located at the community level and, thus, informal community-based organisations within civil society (Norlund et al. 2006:34). Fforde (2008:32-33) describes the informality of groups in 3 provinces of Vietnam, especially in the period of the 1990s and early 2000s.

Scholars have discussed whether group are voluntarily or state-led formed (Benedikter and Waibel 2013:10) and what degree of autonomy from the state they enjoy (Norlund et al. 2006). The very fact that either claim can be reasonably argued demonstrates that groups stand in the space between state management and self-organisation. The informality of groups claimed by Fforde (2008b) and Norlund et al. (2006) is based mainly on the fact that many groups in the 1990s and the 2000s were first set up by the farmers themselves (Fforde 2008b:32-33). However, the authors also assert that local officials support these efforts, and many groups were later recognised by the state and put under the management of the local units of various mass organisations, primarily the Farmer's Union and Women's Union (Norlund et al. 2006:26). These groups have come partly under state influence and management, regardless of who initiated their establishment. Ultimately, groups are part of the Vietnamese system in which the governmental bureaucratic structure penetrates into every area of all administrative levels.

The persistent question is whether the existence of groups indicates state weakness in achieving full control over groups or whether groups, by their nature, emerge from the strategic ground at the interface between the formal state and informal community. The position of groups is analysed by an ethnographical investigation of pumping groups at 24 communes in Can Tho (Thoi Lai, Co Do, Thot Not and Vinh Thanh districts), with in-depth analysis of 3 communes in the Thoi Lai district.

Pumping groups in the researched communes

Pumping groups, as shown in Table 4-7, are formed seasonally for the purpose of collectively pumping fields before sowing the winter–spring rice season. At the end of the flooding season, usually at the beginning of October (lunar calendar) (around November), water starts receding from the canals. Based on experience, farmers then want to start the rice season as soon as possible. Early sowing in November enables a harvest before or in February, the first month of dry season. In rice production, this season is especially important because it offers the highest yield compared to the other 2 seasons (see Table 3-4 in Chapter 3 for the current pattern of triple rice cultivation). Often, flood water from the previous season enters the fields through (unstable) low dykes or over the dyke flood lines (water-rising) season[59]. Artificial or pressurised drainage is required as the season starts when water is still high in the fields. After field drainage, farmers level the land, eliminate yellow apple snails and can start sowing. Thus, field drainage by pumping marks a continuity of human–nature interaction and human adaptation to natural condition based on past interactions and evolution, in this case, the practice of irrigated agriculture and 3 seasons of rice per year.

In most areas of the researched district, field drainage or water pumping-out requires a collective effort. Individual fields separated by ditches and ridges make one compartment, which is covered by a stable dyke system (Picture 4-2). At the planned start of field drainage (October in the lunar calendar), the ridges are completely submerged, along with part of the covered dyke. Water can be pumped out only for the whole compartment, resulting in the interdependence between the farmers whose lands are in the same compartment. 'Compartment'

[59] Delta inhabitants call the flooding season the 'water-rising season', emphasising the value of flooding for the Delta, as well as the damage. Delta inhabitants' perceptions of floods and the beautiful flood are comprehensively analysed by Ehlert's (2011) PhD book: 'Beautiful flood – Environmental knowledge and agrarian change in the Mekong Delta' (Ehlert 2011).

is used as term for the natural boundary of a zone in the management of field drainage, while 'zone' refers to a management unit in which farmers and local cadres organise activity.

Collective pumping with individual diesel-powered pumps becomes necessary to start the season on time and reduce the cost of fuel for pumping. One farmer states:

> Leaking is significant; thus, an individual pump does not work. Only 2 hours after stopping [the pump], water in the field reaches the same level as before pumping. (interview 05.12.2011)

These patterns in pumping groups have been confirmed by researches in areas with the same dyke conditions and cropping pattern, for instance, research on groups in other districts in Can Tho (2009-2010 survey by Benedikter and Waibel 2013:14, Pham Cong Huu 2012), on pumping groups in An Giang province (Howie 2011) and farming system in the Mekong Delta in general (Nguyen Duy Can et al. 2007, Nguyen Ngoc De 2006). After years of interacting with natural and physical conditions, farmers and local cadres (most of whom are farmers) know and understand the necessity of pumping as groups.

Picture 4-2: Corner of the compartment after the pumping-out (Huynh Linh, Dong Thuan commune, 05.12.2011)

This land area with a banana plantation is a covered, stable dyke[60] of a compartment used as a road. In the compartment, the ridges separating the fields were submerged before the field drainage. Pumps set along the covered dyke drain the water from the whole compartment.

The need for collective action or co-operation by farmers varies among areas and depends on the existing infrastructure and natural conditions of each area. Given the internal hydro-ecological diversity within local units, even the same hamlet or commune might have different forms of field drainage. In addition to collective field drainage, farmers with stable dykes protecting their fields can practice individual drainage. Many apply the rice–fish model in which the dyke

60 In this area, 'stable dyke' refers to the August dyke system covering the rice cultivation land. An August dyke is distinguished from a high dyke (which is more popular in An Giang) as it is lower and partly submerged during the flooding season (for more on the dyke system, see Pham Cong Huu 2011).

cover is strengthened for aquaculture during the summer–autumn season. To a great extent, these farmers gain independence from others in deciding when to pump water, which is one factor in farmers' co-operation analysed in Chapter 5. The last and most recent form is contracted pumping in zones, in which companies or pump owner(s) take over field drainage based on individual contracts with farmers. This form involving large-capacity pumps is seen as a modern, market-influenced strategy driven by the growth of the private business sector and the shortage of rural labour.

Farmers in the same compartment perform field drainage as a group (*bơm theo tổ*). Local cadres join collective field drainage in the making of pumping groups. Thus, pumping groups operate seasonally — once a year — and conduct one activity: pumping water from fields for winter–spring rice sowing. The groups are only formed or gathered (again) immediately before the activity in October or November (lunar calendar). The covered area and the number of members in a group vary; the average is 20 households and 20 ha. Although the position of canals and fields play a decisive role in making the boundary of a group's responsibilities, groups often do not expand beyond hamlet boundaries because of the matching between hydrological (canal) boundaries and the administrative set-up. The clear alignment of the administrative, social and environmental organisation of groups is another illustration of the long process in which stakeholders have interacted and adapted to nature and the available physical infrastructure.

Farmers within a group do not necessarily come from the same hamlet or commune. In fact, given inheritance practices and the process of land accumulation over the years, a farmer might have land in different places, even different administrative units. Pumping groups are seasonal, so a group's covered area and membership might vary between years. Amongst other various factors, dyke conditions and water levels are the most important. The covered area is formed annually from adjacent fields by the farmers who, to a certain extent, are economically and socially interdependent. Pumping groups also contract and expand based on the contingencies of participation with both social and ecological dimensions, which are analysed in Chapter 5 through the ethnographic study of the communes' field drainage activities.

Evolution and institutionalisation of groups

Viewing pumping groups as the legalisation of the social organisation of farmers is insufficient. Given the context of the Vietnamese social context of which the state is part, groups are evolving and being institutionalised in a nexus between voluntary activities and state management. This process is illustrated by the diverse narratives of group formation and changes, revealing the interrelation between the state agenda and group and local initiatives.

Emergent groups

Emergent groups are formed by farmers who have a common interest in pumping out water or face the common threat of floodwaters delaying rice cultivation. The commune state later recognises and gives these groups a status. ĐH hamlet in the Dong Thuan commune is representative of this type. In the 2011 pumping season, the hamlet had 4 pumping groups: Groups 1, 2, 3 and 4. Nineteen interviews, observations of 6 meetings and several field walks reveal the history of groups' evolution. Hamlet and group leaders all confirm that farmers initiated the establishment of all 4 groups. A group leader states:

> Some time ago, when water was in late recession, we asked others to come together and pump [water out of the field]. From the upper part [higher area], we just looked for the high enough levees from the zone. … From the success in doing that, we continue until today. From the beginning, we just did [starting the pumping process] without waiting for the commune [commune's state]. We started from the upper zones. I took responsibility without the commune. (Interview 05.12.2011)

In this hamlet, about 7 years ago, when the high water in the fields posed a clear risk that presented an essential need for collective pumping, 2 experienced farmers took the initiative to form Group 2. Others then copied this model to form the other 3 groups. Thus, from the beginning, the groups were formed voluntarily and operated autonomously. Not long after initiation at the grassroots, the action was officially acknowledged by the hamlet cadres, most of whom are the members of at least one group. Subsequently, the actions were reported to the CPCs (CPC). The groups continued to operate without any official status for 2 years until the CPC issued a decision to recognise them (interview 05.12.2011). Although commune cadres could not locate the original documents, the decisions issued for the 4 groups in 2007 most likely correlated with enactment of Decree 151 on co-operative groups. In addition to performing field drainage, the group leaders in this hamlet have since acted as the information gatekeepers for each zone and the agents who transfer state policies and programmes to farmers when requested, particularly the sowing schedule which is closely related to the pumping schedule.

Making such groups official is perceived and indeed claimed by the state cadres as an action to '*capture the inhabitants' wish*' (state cadre, interview 29.12.2011). Above all, the legalisation of farmer's initiative in this case resulted from the fact that the groups also serve the interests and purpose of the local state agencies (Benedikter and Waibel 2013). Discussions with the local cadres and farmers show that, with official status, the groups have evolved but not become a rigid, state-designed structure. Rather, groups coexist with bureaucratic procedures through the regular presence of hamlet cadres, the holding of meetings and operation of pumping groups and continue to follow their own arrangements institutionalised from the influence of both local arrangements and state interference.

Symbolic groups

Coexisting with and sometimes mistaken as pumping groups, some groups whose their operation had been terminated are still mentioned in the local planning of rice production. For instance, loan-getting groups formed in the 1990s today exist only on record and engage in no activity. Xuan Thang, which in 1996–1997 encompassed a larger area than the present-day commune established in 2003, was divided into groups under the state's guidance (former commune cadres, interview 18.10.2011). Except for several groups which later developed into production groups (e.g. the orchard group, the animal husbandry group), the majority of groups was restricted to engaging in loan-getting. In the 2011 field research, only some former local cadres identified the name 'loan-getting group' in the later period. Previously, they had been called co-operative groups, creating confusion about their features and conditions.

In the 1990s, the state guided the formation of groups intended to serve only the purpose of obtaining a loan from the agriculture bank (former commune leader, interview 21.12.2011)[61]. What happening was indeed part of the overall picture of when '*banking reforms and the breakdown of rural credit co-operatives led to the establishment of numerous saving groups*' (Seibel 1992: 64, 77f., cited in Waibel and Benedikter 2013:9-10). At the time, farmers needing a loan had to be in a group as banks did not give loans to individual farmers. Instead, the bank loan was given through a facilitator — the group's leader — who was also the guarantor

[61] In credit programme offering loan to farmers without a mortgage and loans by prestige (*vay tín chấp*), the bank gave loans only to individuals in a group represented by a prestigious person. The bank makes transaction only with the group representative. These loans types continued through the late 1990s and peaked only in 1998, with about 21% of the targeted population. The Mekong Delta, in particular, had received more loans of this type than other regions (Deininger and Jin 2008, cited in Kirk and Nguyen Do Anh Tuan 2009:17). Recently, the Vietnam Bank for Social Policies joined the credit scheme and became more dominant than the agricultural bank in loan programmes for poor households and farmers.

for the loan without mortgage[62] (hamlet cadre, interview 22.12.2011). Forming numerous loan-getting groups was a local strategy agreed upon by the local state and farmers in order to maximise the benefit from the state loan programme during the 1990s. Some later become groups for information sharing or delivery of the sowing schedule. Groups of this type are fluid, existing only nominally at times but, at other times, coalescing to respond to various incentives or challenges.

From their start, the loan-getting groups were very different than a typical co-operative or production group. A group leader states:

> It was not there [in the group] the sharing, but only the benefit of getting a loan from the bank to invest in production to increase household income. At that time, everything was individual, from sowing to harvesting, except a form a labour exchange during harvesting [vạn dần đổi công[63]]. (interview 22.12.2011)

Co-operation in this type of group did not involve sharing. The group only acted as a tool to benefit individual members. To a certain extent, one may question whether these entities should be considered as groups at all during their dormant period between various activities. When the reason for groups' existence disappears, they become dormant until a new purpose for them arises. Since the 1990s, a purpose has emerged from demand for organised co-operation in production, particularly triple cropping which requires collective field drainage. In this context, some loan-getting groups, with the encouragement of local cadres, have been transformed into production groups, or, in other words, been reactivated for production activities. However, the number of transformed groups is few, leaving the rest dormant. As well, the groups' area and membership changed significantly over the course of negotiating co-operation. The diversity of groups would not have become a problem if there were no confusion about groups' status which affects the local planning for production. Local cadres desire to effectively organise local co-operation based on dormant groups, resulting in the failure of group organisation and the waste of time and effort to improve unfeasible groups.

---

[62] The same mechanism has been applied in the recent years by the policy and social bank. Recently, the agricultural and rural development bank has moved to individual loans with mortgages.

[63] '*Vạn dần đổi công* (VDDC) is a common form of labour exchange which remains active in many areas. Young farmers in the same commune join together in a VDDC group, whose members help harvesting each other fields. Labour is collected, and work is done faster. When group members finish harvesting their farms, their labour service might be given on demand to other farmers.

Ultimately, local cadres accept local arrangements of smaller, completely different units for organising production co-operation. These units can be seen as the gradual formation of the next generation of groups in the hamlets.

Non-groups

The third type of group reported in Truong Xuan B commune is the absence of groups in the commune. Despite a similar form of organising collective field drainage, both farmers and local cadres do not recognise these entities as groups; rather, they are simply organised zones for the common purpose of pumping water. While field drainage in this commune, as same as in the other 2 communes, is influenced by both state policies and local arrangements, this case illustrates several issues: state cadres' misconceptions of groups, the continuity of groups' operation and the coexistence of farmers' objection to group identity.

Through the 1990s, a form of self-management/security group under state command was established in the Delta. Each hamlet was divided into groups, each with their own leader. The leader was responsible for tracking migration from and immigration to the area and reporting to local cadres any security problems in the area, mostly related to politics and social safety. The groups stopped operating sometime in the late 1990s as their security tasks were shifted to state agencies and police stations. However, recently commune cadres have (deliberately) mistaken the old security groups as production groups, or the pumping groups of the commune. However, beyond the occasional mention in conversations, the security groups have no role in today arrangement of local production and internal affairs.

Although the commune gives no official recognition to production or pumping groups, these use exactly the same arrangements for communal co-operation in field drainage activity. The present coexistence of disproval of the term 'group' (as part of the Communist collective economic system) and the overlapping boundary between present pumping zones and old co-operatives (during collectivisation in the 1970s) is, indeed, a result of gradual interaction and accomodation in the past. At the hamlet level, the following statement is common:

> At present, there is no group; there is no kind of organisation in the hamlet. Rather, farmers work together among themselves. Before, we had co-operatives [tập đoàn sản xuất], but they are all dissolved. (Hamlet cadre, interview 14.12.2011)

However, farmers organise the collective field drainage in zones with the guidance and management of the hamlet cadres. Observations of the field drainage organisation in Truong Thuan hamlet demonstrates that all 4 pumping zones initially followed the boundaries of the 4 co-operatives which existed in the 1980s. There is continuity in local arrangements which results from the unchanged basis of relying on the canal structure to establish zones. Although the old arrangement is reused, the old identity is denied. As stated in the previous quotation and 31.11.2011 interview, farmers fear that using the old name or a similar one (e.g. 'group') could lead to the re-emergence of the old command-and-control system which they rejected. Therefore, in this commune, groups exist but without official status, and the groups in this commune are indifferent to the others in the interaction between state intervention and other local arrangements.

Discussion of group evolution and the state's role

Based on the elaboration of these 3 cases, I conclude that these groups have diverse formations and continue to evolve through a process of institutionalisation. This evolution is based on human interaction with the natural conditions, existing infrastructure and current socio-economic and political situation.

Group evolution depends on natural conditions (climatic and water regime), available infrastructure, sentiment toward state policies and co-operation among farmers. Thus, groups are not static but evolve because of various incentives. In this case, with the changing incentives and state control, dynamism appears to make diversity serve the selection of the best practice for a certain period. Thus, the institutional process is strongly influenced by the interaction and negotiation among various stakeholders, here, between the policy-level state and the local society. Selecting certain arrangements and disproving others is part of the legitimation process that decides the survival of institutions at certain periods of time.

Regardless of who initiates a group, other stakeholders will become involved, confirming groups' position in the space between bureaucratic structure and local arrangements. Groups and their characteristics are institutionalised over time under the influence of both the official structure and local design. Combined with the observations of the 3 communes, this research suggests to the erasure of the boundary between formal and informal feature of groups. It is impossible to categorise a group as official or informal. Rather, all groups stand at the interface of state policies and local interests, affected by various incentives for the different actors: local states' assistance (hamlet People's Board), mass organisations and farmers. The diversity and mediating role of groups have both been features of Delta life and a constant challenge in local governance.

Putting emergent groups under local state management, guiding the establishment of loan-getting groups and efforts to legalise groups are proof of state interest in group. Digging beneath the stated purpose of achieving national goals, what drives the state to encourage the establishment of groups, to support them and to remain involved in their management? The state interest in penetrating local affairs, which is further analysed in Chapter 5, derives from 2 goals: to maximise the chance of success and/or the completion of task in local production and to maintain the image of '*the state cares for the people*' and state control over every corner of the locality. Through groups, local state can mobilise the inhabitants to implement policies and achieve the allotted agricultural and economic targets/mandates. Simultaneously, groups are a form of community gathering that can create certain political risks for the Vietnamese state and so need to be closely managed and controlled. While the production task is publicly confirmed by all interviewees who are state cadres (including mass organisations) in the 24 communes, the state control ideology is only mentioned in internal occasions.

Local cadres have the responsibility not only to establish and manage groups but also to nurture and consolidate them. However, the state's role in group management remains unclear due to the confusion in responsibility for groups amongst the state agencies and the uncertainty of state influence on the effectiveness of groups. Local state incentive for supporting and managing groups depends on the agenda and leadership of the locality. In case of loan-getting groups in the Xuan Thang commune, the CPC has assessed the groups in the commune as 'degraded' and in need of strengthening (commune meeting 09.11.2011). The term 'degraded' implies that the groups performed better in the past and illustrates the belief in the importance of the state's role in guiding and managing groups.

Blame for the groups' degradation is passed from one party to another because of confusion over task responsibility. While the CPC blames the hamlet cadres' and Farmer's Union lack of ability to manage groups (interview 21.12.2011), a group leader asserts that:

> I can say that groups are weaker partly because the commune [CPC] shows less concern. I remember that, during the regime of Mr. XYZ [former leader of the commune], all groups' leaders were invited for the production planning meeting at the commune. Then, CPC cadres, hamlet's cadres and groups were all together going for field observation …. Because the group leaders have neither salary nor allowance, the hamlet people's board was flexible in giving in-kind benefits for the group leaders, for example, to be on the priority list for becoming a model in a production project or for a loan programme. (Interview 29.11.2011)

According to the hamlet cadres and group leaders, the CPC has a driving influence on the effectiveness of groups. In another conversation, a Farmer's Union cadre claims:

> It should be the CPC which establishes [groups], cadres at the agriculture sector [almost all CPC cadres], the Farmer's Union in the commune and hamlet, and the hamlet Party unit to nurture it. However, until now, there has been no clear division at all. (interview 21.12.2011)

The confusion in responsibility stems from the unclear task division in group management at the local level, similar to on the provincial and district levels where agencies point to each other as responsible(the case of Farmer's Union, Department of Rural development and the Co-operative Alliance) (interview 27.06.2011, 28.06.2011, 01.02.2012). The uncertainty in task division and the complexity of the sector (groups are neither fully under state control nor free from the state structure) have prevented the state from further contributing to the groups' improvement. However, the state generally has accomplished its goal by maintaining nominal control over these groups, either by initiating their formation or keeping record of them and issuing decision papers for them. The state's persistent attempt to have minimal control over groups contributes to the evidence that state incentives have penetrated into every corner of society which serves both management and security control targets (avoiding and proactively responding to any potential rural unrest).

From farmers' perspective, they accept the state requirement to be involved and approve of state support for conflict resolution, but they do not expect state involvement to entail close monitoring. As recounted many times, the farmers expect state cadres to care of and help groups, especially in providing seeds for rice cultivation and solving conflicts during field drainage. For instance, the issuance of a group establishment decision under Decree 151, despite its uncertain effect on groups' operation, is at least considered by farmers as '*a sign for the farmers to know that the state is there to support them*' (a group leader, interview 30.12.2011). The nature of group organisation also creates a rationale for state involvement. As mentioned in the interview 29.11.2011, the state's influence is related to group leaders' incentive to manage groups Leaders and other individuals' incentive to lead and tolerate free-riding actions are key to the survival of present-day collective actions in rural areas (see Chapter 5 for details of the organisation of field drainage). Working for groups remains voluntary, resulting in an unstable allowance based on household contributions (ranging from 1,000-3,000 VND per *cong-1.300sq m*). Thus, it is often complained that:

> The group leaders have no allowance. They are eating at home, wearing clothes made by their wife, and working for the group. (Interview 29.11.2011)

In this situation, some state's offering of in-kind benefits motivates group leaders to continue working for groups.

Over the years, farmers have grown accustomed to the presence of the state in all matters of their lives. In other words, the farmers and the state have become interdependent. However, state intervention manifested as the legalisation of groups by the issuance of decisions, so far has played no practical role in the functioning of groups. The question is whether this absence of impact stems from the state's lack of capacity to influence groups or from local resistance favouring the individual farming system. Chapter 5 continues to explore this issue and prove that the absence of impact is the result of contemporary interactions between the state agencies and farmers in which both have various objectives: The state wants to keep the groups in control while lessening its financial and management burden; the farmers want to make decisions about on their own fields and want the state to give support, when needed.

In sum, groups remain a complex, diverse sector in which state engagement is unclear. However, the state's presence in groups' formation and operation is clear. Farmers are becoming familiar with state involvement, and some even believe in the influence of state support on groups' wellness. The extent of the state's influence on groups remains dependent on the specific cases in which the local state plays an important role. In general, one can conclude that contemporary groups, by nature, are both dependent on the state and autonomous from it. Their special position and relation to the state, especially when they are considered to fall within the scope of civil society, confirm Weller's (2005) statement: '*The relationship between civil society agents and governments are rather "symbiotic than antagonistic*"'[x] (cited in Waibel 2013). In the context of groups, this relationship not only denies the idea of a linear, oppositional force against civil society which confirms the work of Waibel (2013:12) but also confirms that the Vietnamese structure of civil society is characterised by the coexistence of state presence in every corner of life and the autonomy of socio-economic organisations. The question is whether issues are solved by using the state structure, informal means or a combination of both in groups. Ultimately, all that matters is accomplishing the goals of the community, i.e. loan-getting, pumping, and harvesting.

Conclusions

The local governance of agriculture and irrigation management in the Mekong Delta is the joint work of decentralised state agencies and other stakeholders, including semi-state agencies and farmers' organisations. There exists a complex interdependence and inter-influence amongst the stakeholders which, under natural, physical and socio-economic changes, drives the evolution of state management and the characteristics of the other agencies over time. While the combination of a de-concentrated structure and local centralisation illustrates persistent structural barriers to sharing power and benefits, the interaction amongst various stakeholders creates local dynamics which can both pose obstacles to state control and complement state inconsistency and uncertainty. Overall, the involvement of local movers is simply part of the Vietnamese structure and the negotiation and co-adaptation processes from which arrangements are introduced, evaluated and legitimated.

Today, the state Party has a 4-level administrative structure characterised by a system of double subordination. Decentralisation in Vietnam initially was a solution to the failure of the centralised state management system which allowed the state to retreat from social and economic activities amid the rise of the market economy and the private sector — namely socialisation and privatisation — and to promote social democracy (Grassroots Democracy Ordinance). During the de-concentration of state administration and authority over budget management, significant steps have been taken to give more authority and autonomy in decision making to the provincial government led by People's Committee. However, researches, including the present study, find that fiscal decentralisation has stopped at the provincial level, leaving a centralised system at the district and commune levels. When necessary, the state might yield specialty functions in agriculture and irrigation, if doing so does not undermine its political monopoly. However, the state knows from past experience that many small changes which do not directly undermine the state can accumulate and, in combination and in large numbers, undermine the state. This concern explains the partial decentralisation and the persistence of the state's dense structure of agencies down to the grassroots level as a means to check potential resentments and antagonisms.

As seen in the irrigation sector, retaining authority, especially over budget management, at the central and provincial levels entails both keeping the official (and unofficial) benefits to those superior agencies and distrusting local governments' capacity to effectively handle authority in local affairs. The case of the irrigation sector exposes the belief and fear that de-concentration and devolution can lead to the loss of power of high modernism and, thus, of the state's legitimacy (Scott 1998). In addition to building and possessing knowledge and technologies as a legitimate power of the (upper-level) state, finances have increasingly become modernism's main source of power in contemporary agriculture and irrigation management influenced by mechanisation and the market economy.

The state bureaucratic system down to the hamlet level is supported by the legal framework and the application of infrastructure power (Göbel 2011:177). While the state gains power from regulating society through institutions and organisations and imposing the standardisation and simplification of society (Scott 1998), the application of infrastructure power also involves state embeddedness in society. At most local levels, embeddedness enables the state to react adequately to local demands, increasing its legitimacy (Mann 1984, ibid:185), and to take coercion action when local tensions become threatening. The state must decide on the extent of embeddedness as too much embeddedness costs resources, while too little risks the loss of legitimacy. One of the state's most hands-off ways of remaining embedded is the recognition of flexibility at the local level through using the state budget (the case of budget for canal dredging at the district level is explored in this chapter) and flexibly applying local arrangements in implementing state mandates and policies (this topic is analysed in Chapter 5 on field drainage activity).

In addition, the Vietnamese state has exercised discursive power — the power to make people want what the government wants them to (Luke 1975/2005) — through the establishment and operation of agencies doing mobilisation work, in particular, extension offices and mass organisations. Through these organisations, the state hopes to impose social order by creating and disseminating discourses which define what is true, beautiful, moral, fair and legitimate (Scott 1987:38). The state, thus, '*build[s] a symbolic climate that prevents subordinate classes from thinking their way free*' (ibid). Discursive power gives the state has lower-cost options to address social problems and regime challenges than repressing or cracking down on opponents (Göbel 2011). As it practices both infrastructure and discursive power, the Vietnamese state can be categorised as mobilisational corporatist state (Kerkvliet 2001b, Kerkvliet 2003). However, with dual accountability to both party–state and local society, the extension office and mass organisations — or discursive power machines — both mobilise local compliance to state policies and transfer local demands which can result in adjustment and adaptation by the state. Similarly, during the process of embeddedness, other agencies at the local level are also influenced by local arrangements and gradually absorb, report to and flexibly adjust the state system to be suitable for the local situation. Therefore, in the Vietnamese context, the practice of state discursive power is likely to produce the dialogue form of the state (Kerkvliet 2001, 2003). Empirical evidences presented in Chapter 5 support this claim.

The case of irrigation highlights that local adjustment contradicts but also sometimes complements the rigid, bureaucratic and centralised state structure at the local levels. Although the inconsistency and uncertainty of the budget for canal dredging motivates local strategies for completing the maximum number of canals in communes, unclear task division and budget allocation increase the scope for flexibility in accommodating the deficit. The management of irrigation also has many points of access which ease adjustment due to the relation to many other issues, especially production. The Vietnamese system of management constitutes a state bureaucratic structure of mandate-giving, decentralised and centralised administration and budget management and local, dynamic arrangements flexibility induced and supported by the state structure.

The state's adjustment to decentralisation and privatisation and recognition of local deviation does not necessarily imply state weakness in command-and-control management nor proves local success in exerting political leverage through resistance. In other words, the policy changes in the contemporary period do not stem from the acquisition of power from either state power or local resistance. Instead, the policy changes can be observed and analysed in the realm of institutionalism in which an arrangement is considered, selected and legitimated over a process of interaction and negotiation. Nevertheless, the institutional process is not a straightforward or organised event of negotiation but includes various forms of interactions, from official to informal communication. The changes often take time and happen silently without producing any clear winner or loser. In some ways, this pattern marks the continuity of the interaction during the Doi Moi in the 1980s — reforming without losing face (Rama 2008). In addition, all stakeholder(s) accept one arrangement from the other(s), revealing the interdependence of those in the same society. These characteristics are empirically illustrated in chapter 5 on the negotiation of local irrigation practices.

Societal dynamics are guided by other stakeholders, as well as the state; therefore, it is important to identify these stakeholders in order to analyse local irrigation management. Figure 5-14 in Chapter 5 lists the stakeholders directly involved in field drainage before the main rice season. In addition to state agencies and cadres, others stakeholders, such as farmers, farmers' groups and private firms, are separate and autonomous from the state management bureaucracy. Joining the practices and being masters in the field are farmers and their organisations. Farmers' organisations, in this context focusing on groups as the main stakeholders in production and irrigation management at the local level, are very diverse, simultaneously autonomous and reliant on the state. Groups are formed under the influence of natural conditions, the available infrastructure and social and administrative purposes; consequently, their evolution has social and ecological dimensions. Groups are diverse in name and activities, and their names do not necessarily reflect their nature. Although groups should be voluntary and autonomous, the state, as it penetrates every corner of society, has tried to subject groups to bureaucratic management. However, given their non-uniformity and the confusing state management structure, groups remain autonomous yet state-supported organisations.

CHAPTER 5

NEGOTIATING THE INTERFACE OF IRRIGATION MANAGEMENT AT THE LOCAL LEVEL

Irrigation or, more generally, water control historically has been a critical aspect of exploiting and settling the Mekong Delta and continues to be the crux of relationships among state planners, local cadres and individual farmers in modern times. With a decline in the capacity for command-and-control agriculture, the state has taken on the role of local animator–expert, providing a platform for discussion and technical suggestions for harvest optimisation. Beyond direct interactions with the state, such as the dissemination of the sowing schedule, local cadres and individual farmers are left to work out activities, such as dyke maintenance, collective field drainage and field canal dredging, within a framework of mutual economic interdependence. Although a markedly capitalist and individualistic orientation to farming has become predominant, social co-operation remains due to the unique hydrological conditions of the Delta. This co-operation has arisen in a nexus of modernisation programmes and the fluid immigration regime of the Delta, in which social openness and flexibility have provided the basis for local people to deal with the ecological challenges of the Delta.

This chapter sheds light on the state-facilitated processes of social co-operation that enable Delta rice farmers to achieve 3 rice harvests per year by exploiting the interface between state planners, local cadres and farmers. First, I look at the dissemination of the sowing schedule, which is intimately tied to the timing of flood recession and farming tradition. Second, I consider the co-operative arrangements for dyke maintenance during the implementation of the sowing schedule through field drainage by pumps and analyse why relative successes and failures are achieved. Third, I examine the shared efforts at canal dredging, the negotiation of which necessarily draws together resources from peoples dispersed over large areas. All of these activities demand co-operation for success, which compels the state to intervene in various ways to facilitate and encourage platforms and incentives for collaboration. I argue that, although the primary support structure of mutual assistance is the ecological conditions that generate economic interdependence, the state has a critical role in overcoming initial barriers to co-operation. Furthermore, while active field drainage and dyke maintenance are mandatory and vital for nearly all farmers and the state to achieve their various goals, canal dredging and widening are less critical, which discourages strict co-operative arrangements and leads to more ad-hoc strategies at the local level. The diversity apparent in these activities enables a broader analysis of the local dynamics of the irrigation management and the intersection of state management and social organisation in the Delta.

Negotiating field drainage and sowing schedule

Rice production continues to be the main farming activity and income source for farmers in Can Tho and the Mekong Delta. Rice production also lies at the nexus of power among individual households, collective work arrangements, state monitoring, technical know-how and intervention by experts. Indeed, rice production not only concerns livelihoods; it also entails rehearsing and reproducing local arrangements, updating skills and negotiating with authorities and fellow farmers. Raising or maintaining production through various practices is the most important consideration for all parties to fulfil their various goals. Of these, achieving reliable water control and yields of a certain amount are the most visible targets. In this section, I focus on the former issue, specifically the expectations and nexus of relations clustered around water control for the very important winter–spring rice season.

Agricultural production was officially returned to households under Doi Moi, but walking through hamlets, talking to farmers, it is not difficult to see the presence of the Vietnamese state policy and structure in everyday conversations and in community activities. Overall, farming in the Mekong Delta and Vietnam, in general, has remained an individual business since the return of individual farming following the de-collectivisation of agricultural production in the late 1980s. However, individual farming coexists with the people's ownership under the state management of all national natural resources[64], as stated in the Constitution:

> Land, water resources, mineral resources, the resources from the ocean, land and sky territory, other natural resources and all properties that are invested and managed by the government are public properties, under the ownership of the

---

[64] In Vietnam, land, as other natural resources, is owned by all inhabitants and managed by the state. Each household or organisation receives the right the use an area of land for certain period of time — 50 years according to Land Law 2003 – 13/2003/QH11. Vietnam faced a problem in the Land Law in 2013 when many grants for use of the land expired. This problem was solved by Resolution 49/2013/QH13, which replaced by the Land Law (amended) /2013/QH13 (approved by National Assembly in November 2013) and regulated the expansion of the usage period for various kind of land use.

> entire population, represented and managed by the government. (National Assembly 2013, article 53, Chapter 3, translated by author)

Indeed, as the manager, the state extends its presence into all corners of society through the dense hierarchical system (see Chapter 4) but to different extents and in various forms throughout the stages of farming process. As manager, the state is responsible for production planning, including for the sowing schedule, farming season and implementation process.

However, there is not necessarily a correlation of the state's presence and its strong influence or controlling status with individual life and social culture. Whether such a correlation exists is a central question of this research. Using field drainage as the object of analysis, I argue that individual households make farming decisions in interaction with other stakeholders, including various state cadres. The decision-making process is built on negotiation between human beings and adaptation to non-human factors in a locality. Thus, the implementation of farming policies and practices are context dependent. This interdependence creates a space or interface in which clashes, negotiation and adaptation take place between stakeholders at the practice level, especially between state mandates and local arrangements.

The analysis of field drainage for the 2011–2012 winter–spring rice season starts with an introduction to rice farming practices, particularly the relation of sowing and field drainage by pumping-out and the general technical aspects of the field drainage. Next described is the state-designed procedure from planning to implementing the sowing schedule as part of production planning. This process consists of steps to transfer mandates from the province to the commune. The process from policy to practice is analysed from farmers' perspective. The analysis focuses on the negotiation and adaptation of participants, including local state cadres, farmers and private pump owners, at all steps of the pumping process. This anthropological analysis, along with narratives from farmers and local cadres, reveals the nature of negotiation throughout field drainage activity. This analysis provides the background for defining the state's position and the local dynamics of interaction and negotiation at the interface of irrigation activity.

Farming practices and field drainage before the sowing of the winter–spring rice season

Rice farming in the researched area and field drainage

The researched activity of field drainage or water pumping-out before the sowing of the winter–spring rice season lies within not only the field of irrigation management but also agricultural production, which is guided by natural, physical and socio-economic conditions and the state's agenda for food security and production planning. As mentioned, rice is cultivated in the alluvial land of the researched area for 3 seasons a year (Figure 5-9). As rice production moves towards irrigated agriculture with water control, such issues as choice for the rice variety cultivated, water control, timing and pest control become priorities. In water control in the Mekong Delta, field drainage to enable sowing in time for the rice season is as important as ensuring a sufficient amount of water for irrigation. Thus, to characterise the sowing period according to natural, physical and social conditions and the state's plan, it is necessary to identify the timing of field drainage. The 2 activities, field drainage and sowing are strongly related, so the state procedure for implementing the production plan and the sowing schedule directly contribute to the imperative for field drainage.

Figure 5-9: Cultivation lunar calendar for 3 seasons of rice per year in Can Tho

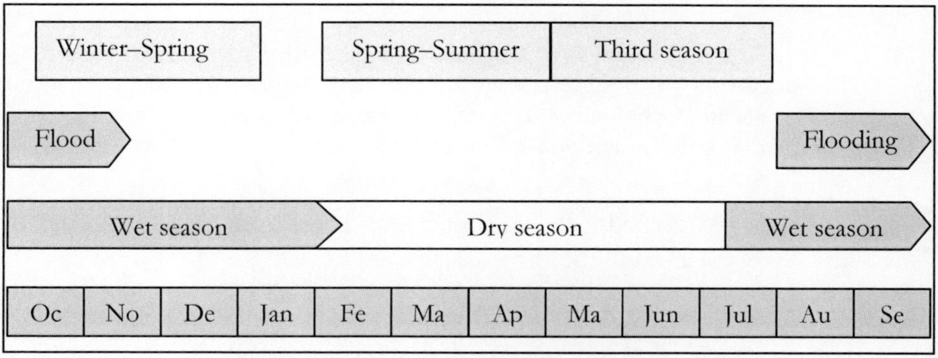

*Note: In general, the Mekong Delta's year is divided into a 6-month dry season and 6-month wet season. Often, most precipitation falls from May to November or December, with a regular dry spell in July. The winter–spring rice season is begun as soon as the flood waters start to recede, usually around October. After the harvesting of the winter–spring rice season, the spring–summer rice season is also immediately without soil preparation (sạ chay — dry sowing) in order to save time and enable cultivation of the third season.

With various options for cropping patterns on rice land (Table 3-3, Chapter 3), most farmers in the Thoi Lai district cultivate 3 seasons of rice per year: winter–spring, spring–summer and summer–autumn (or third season). Amongst these, the winter–spring season has been proven to be the main and best season for the highest yield, 6–7 tonnes/ha compared to 4–5 tonnes/ha for the summer–autumn season and less than 3–4 tonnes/ha for the third season (2000–2010 data on the rice yield in Can Tho, Statistical office of Can Tho city 2008, Statistical office of Can Tho city 2011). One farmer states '*there is no way to lose during the winter–spring season*' as the climate in this season is the most favourable for rice, and '*if you do* [lose] *for the winter–spring season, you will be in debt for 3 years*' (interview 26.10.2011). Consequently, the season has gained much attention from farmers, their organisations and the state, making it a representative time in which to study the interests and activities of rice stakeholders.

Traditionally, the winter–spring season starts immediately the recession of floodwaters in order to enable the best harvest and timing for triple rice cultivation. In farmers' experience, the floods start to recede in late September in the lunar calendar (often October), and the first low water (*nước kém*) of October (lunar calendar), typically 6–12 October, is the best time to start sowing so that the current rice varieties, with an average growing period of 90 days, will enjoy the best weather of the year from October to January (lunar calendar). A well-known rice expert in the Delta states that:

> The yield of the winter–spring rice is often higher than other seasons in the year. Besides the factor of higher soil fertility (due to the supply over flooding season), higher radiation, the higher temperature amplitude between day and night [of the period] are also important to explain the higher yield. (Nguyen Ngoc De 2006:78)

Farmers in the researched area confirmed that climatically, it is not good to prolong the harvest until the start of the sunny/dry season (beginning of February in the lunar calendar) (group interview 11.11.2011). In addition, with sowing done in early October (lunar calendar), the field can be harvested in early January (lunar calendar) (on or after the Tet holiday[65]) to secure a large amount of income for the coming year. Traditionally, such income is also a good sign for domestic prosperity. On a practical level, starting the first rice season early makes the third rice season possible.

October sowing has become a tradition and piece of knowledge that farmers teach their children. The tradition was established with the International Rice Research Institute's (IRRI)[66] introduction of the miracle rice IR8, which first enabled the double season in small areas in the late 1960s, was expanded

---

[65] The Vietnamese new year holiday.

[66] The IRRI is an independent, international research and training organisation which aims to reduce poverty and hunger, improve the health of rice farmers and consumers and ensure the environmental sustainability of rice farming.

especially during the collectivisation period (1970s–1980s) and continued with more high-yield varieties in the 1990s. Under the influence of nature, the available infrastructure, government interventions and social dynamics hiding collaboration and competition for prestige, farmers have considered, selected and adapted the rice-based farming system to achieve controlled irrigation and 3 seasons per year. The traditional practice still survives as it is still valued by relevant social institutions. For instance, the tradition of organising wedding ceremonies in households at the beginning of the year (January in the lunar calendar), which stems from the regular profits of rice harvesting, makes the large, stable income from the winter–spring rice a necessity to realise the family's plan. Furthermore, a high yield and stable household income are also the targets of the state in production, food security and social prosperity.

Despite the desire to start the season on time, the decision-making process for the sowing date depends on a number of factors: water level, infrastructure (dyke system), state sowing schedule, social acceptance and farmers' ability and willingness to act collectively in the field drainage. In Thoi Lai, as in most of the Delta's rice land where fields receive water during the flooding season, cultivating the winter–spring rice season consists of several basic steps: field drainage, dyke enforcement before and after water pumping-out, field levelling, yellow-apple-snails elimination[67], sowing, growing time and harvesting. Field drainage by pumping is not necessary in certain areas where high dykes in mostly closed irrigation systems cover the rice fields, resulting in very little inundation during floods. Pressurised field drainage is not applied in inundated zones where gravity drainage is possible due to a lower water level in the canal. Outside these conditions, an inundated area with high water levels in both fields and canals and an unstable dyke system requires collective action in pumping. The actions are made possible by the availability of individual pumps and an agreed-upon arrangement among economically interdependent farmers.

Consequently, field drainage happens at the end of flooding season, often in October (lunar calendar) when water is low enough. When the water in canal is lower than covered dyke, the decision is made to drain water from the fields to start preparing the land before sowing. Unless gravity drainage is possible, pumping is necessary to remove water from the fields. Additionally, strong

[67] The yellow apple snail has been one of the most harmful pests in rice fields. Without proper elimination before sowing, the snail population can damage many of the young rice plants. A chemical method is more common, while manual collection of snails to use as fish food is also practiced. Recently, rearing and selling the yellow apple snail has emerged as a business in the Mekong Delta.

dykes which prevent water from flowing across fields are missing in many areas of the researched district, so collective field drainage is more common than individual pumping in large zones covered by a stable dyke. The common farming practice for the winter–spring rice season with water pumping-out is illustrated in the Figure 5-10 (observed in Can Tho).

Figure 5-10: Process of field drainage organisation at the commune and hamlet levels (made by author)

Collective action in field drainage can be seen as a response to a common threat, which is the main incentive for farmers to perform communal work. The nature of co-operation in water pumping has been detailed by Howie (2011) based on research in An Giang. In this sense, collective action and co-operation are gained through the process of neighbours working together in building fences (Popkin 1979) or collective action to overcome individual lacks of infrastructure. However, willingly working together is only one facet of human co-operation; many other factors influence the incentives and decisions for communal work. These factors are further analysed in the process of organising collective field drainage.

Technological diversity in field drainage

Collective field drainage entails collecting diesel pumps from households in the same zone to start pumping at the same time. Those individual pumps have a small capacity and are used every day for small-scale pumping and boat transportation on the waterways (Picture 5-3). High-capacity pumps (Picture 5-4) are used in contracted field drainage by one or more individuals (farmers and non-famers) working a large area of land, often 70–100 ha or more (observations in communes in Thoi Lai district in 2011–2012).

Picture 5-3: Individual diesel pump used for collective pumping; Bottom – pump with fuel can (Huynh Linh 2011)

Picture 5-4: High-capacity pumps used in contracted pumping (pumps and pipes) (Subramanian 2011)

The increasing alternatives for field drainage are in line with the on-going state policy to form and expand mechanised collective and contracted agriculture (i.e. programmes to give loans for individual to invest in farming machinery). These alternatives also serve programmes, such as standard rice field, new rural areas, contracted land preparation and contracted sales. Therefore, as with the sowing schedule for production planning, the local state facilitates and monitors the implementation of pumping solutions in order to protect farmers' benefits and, consequently, state production targets. During the negotiation process of both collective and contracted pumping, hamlet cadres often act as the facilitators of meetings and contract establishment as farmers' representatives. According to the hamlet and commune cadres, contracted field drainage is new, so pump owners have to sign one general contract with the CPC and individual contracts with each farmer in the pumping zone. Overall, contracted field drainage contributes to the common sense of a persistent state presence in the rural area. As the trial solution in filling the gaps in current practice, contracted pumping, whether initiated by the local cadres or farmers, is monitored by the local state.

The tendency to shift to a service provider–consumer system in agriculture production contributes to the movement towards a market-influenced or capitalist economy. In this market economy, the world outside the village world has increasing influence on individual households and their decision to pursue either social cohesion and collaboration or individualism — e.g. my-family–first attitude. However, this shift should not be seen as a constraint on or the destruction of social values; rather, in the Mekong Delta, it is another dynamic guiding institutions that have been formed, negotiated and adapted over a complex history of human settlement. When an arrangement is validated, it is accepted and subjected to the process of institutionalisation (Cleaver 2012). Institutions continue to be negotiated, and their legitimacy, which relies on the availability of reasons, is questioned. Next explored is how the current dynamics affect the local dynamics and various stakeholders, leading to contemporary practices.

In addition to collective field drainage with individual diesel pumps, tenants in some special areas, such as collective farms (Song Hau and Co Do in Can Tho), have high-capacity electrical pumps. Electrical pumping stations are becoming a model in the Delta (either under testing or as the continuity of past arrangements[68]). In the context of field drainage, electrical pumping stations

[68] In the past, electrical pumping stations were set up in some pilot areas in districts following provincial policy. However, many did not survive because of financial challenges and human-resource shortages (case of Vinh Thanh, interview 06.07.2011).

and collective field drainage can be seen as the interfaces in which technology, the capital for getting and employing technology, and the knowledge and models which guide the use of technology (from small pumps to high-capacity electrical pumping stations) continuously redefine who, how and how long the collective pumping periods last and who participates, co-operates, wins or loses.

Field drainage: Social organisation of multi-identity individuals in the current context

Field drainage, when it becomes a communal activity, encompasses not only a simple collection of tools but also the complex organisation of human beings to use these tools to achieve certain targets. Stakeholders who have different identities, interests and agendas join to organise and negotiate social arrangements in order to achieve individual and communal benefits.

The field drainage activity can last from 2 weeks to a month, counting from the hamlet pumping meeting until sowing. After collective pumping, fields are separated by local dykes, and farming activities return to an individual basis. This pattern confirms the looseness of community ties in the Southern Delta and the continuity of farmers' collaboration in land preparation, seeding and harvesting (Nguyen Ngoc De 2006, cited by Waibel and Benedikter 2013). The adoption of both collective and individual practices demonstrates the heterogeneous reality of farming practices in this area where all options contain choices resulting from adaptations during the process of learning by doing.

Amongst the stakeholders in field drainage, farmers constitute the majority. From a top-down perspective, they are the final object of the state policies. They are also the masters of the field, so the decision of what to be done for field drainage relates to their rice field. However, farmers do not form a homogenous group and contribute significantly to the diversity in the local level of irrigation and production practices. In the researched district, being a farmer means having a piece of land ranging from several *cong*[69] to several ha (or even more than 100 *cong,* which is approximately 130,000 sq m). The common cropping patterns are either monoculture or a combination of paddy rice, dry crops and orchards. Some farmers close their area with a stable, earthen dyke to culture fish or carry out rice–fish integration model. The significant variations in land size stems from the process of land accumulation over time under

[69] *Cong* is the land unit commonly used in the Mekong Delta and covers an area of 1.300 sq m. The standard *cong* is 1000 sq m and is commonly named *small cong – công tầm nhỏ*.

cultural arrangements, such as inheritance, and socio-economic changes, such as labour migration and land use transfer.

As a result of cultural and market processes, the farmers or land owners in the researched area can be either locals or migrants who moved from another hamlet, commune, district or even another province. In this research, the term 'local' does not refer to the Delta's indigenous inhabitants but, rather, families that have 2–3 generations living and farming in the same area. Amongst the newcomers, the most common reasons for migration are family (mostly women) and the opportunity for a better income (interview 31.11.2011). The heterogeneity of the researched communes' population profile illustrates the typical picture of the rural Mekong Delta, where the historical dynamics and social adaptation have created an open society which welcomes newcomers to settle as part of the existing community (see Chapter 3 for a detailed analysis). This social characteristic complicates the social system and suggests that the Delta people have evolved a strong degree of flexibility. The question is whether farmers' profile can influence individual power in negotiation. This question is considered in the case of negotiating field drainage later in this chapter.

An obvious illustration of the openness of researched area is the interdependence in the agricultural production system of farmers inside the hamlet and outside farmers who also farm land inside the hamlet (*xâm canh*).

> There are quite many xam canh farmers in our hamlet. The group leader has the responsibility to inform them to come to the [pumping] meeting … Those will bring the pumps and start together with all, although they are from other hamlets, communes or other districts. (interview 29.11.2011)

This statement, along with the hamlet observations, shows that *xâm canh* farmers get the same social power through negotiating production processes, although these farmers have limited access to information, especially when collective decisions are negotiated through everyday encounters in the hamlet (e.g. market, rice fields). In terms of social cohesion, one can assume that staying outside the community results in a certain degree of difficulty in accessing production information (e.g. pumping and sowing dates, hiring machines for ploughing, harvesting, canal dredging, making contracted sales). On the other hand, these farmers might have more freedom and independence from the social ties and obligations constructed by living in the same neighbourhood. In fact, with the Mekong Delta open society's free acceptance of outsiders, every farmer, regardless of residence, is equally recognised by others and has the same power in negotiating production matters. This was confirmed in all 3 researched communes, where holding land gives outside farmers the same rights and, thus, a position at the negotiating table for field drainage.

This situation exhibits a fairly capitalistic feature of the farming society. The most important matters in the negotiation process which decides farming practices are the natural and physical conditions and individual choices to co-operate. The limitation caused by living a far distance from the field and so from the negotiation interface is reduced by achievements of modern society, such as mobile phones. However, those who live away from the meeting place and often hold other land in other zones might choose to engage in indirect negotiation through fellow farmers in the pumping zone and tend to follow the decisions made by their field neighbours (recorded in Thoi Hoa B hamlet, commune Xuan Thang). These field neighbours, although they have weaker ties compared to the ideology of rural neighbours, all rely on interdependence, mostly in economic and farming activities.

In addition to being farmers, individuals might hold other responsibilities that significantly affect their view of and behaviour towards social co-operation. Nearly all state cadres or state assistants (commune and hamlet cadres, respectively) are farmers. As state cadres, they facilitate and mobilise tasks in implementing the state policies. The cadres also should be models for fellow farmers, the first to implement and pilot activities encouraged by the state.

State cadres and locally prestigious individuals or social leaders frequently overlap. Some farmers take on the facilitation role in farming collective actions. These are often good farmers (good at farming practices, leading to good yields) and good neighbours (in harmony with and helpful to the neighbours). Consequently, they gain prestige over the years, managing and facilitating certain collective farming practices and, in some cases, the formation of groups or clubs. Most likely, these are former or future local cadres or state affiliates. Although local facilitators are only managers, not leaders who can command other farmers, they do have a role in changing behaviour of the others, as well as organising the timing for collective works and sharing responsibilities. The overlap between the state and local arrangements of leadership, indeed, proves to be a result of the state's understanding of the local practices. Through interactions, the state recognises local leadership arrangements that can be exploited to maximise the success of state management. However, the diversity within the commune in household landholding, profiles, multiple responsibilities and way of participation might make the rural collective actions uncertain.

Influence of farmers' assets in negotiation

The most powerful determinants of negotiating power and behaviour in communal works are economic and social values which are brought to self-negotiation before a decision is made. Before participating in communal works became a form of social cohesion proving one's social responsibility or 'social spirit' (Reis 2013:23), collaborating with farming neighbours began as a strategy to overcome difficulties individuals could not handle alone. Social relations grew along with collective works in the community and became a social tie or requirement acknowledged by participants and influencing human behaviour — or the making of farming institutions. However, the social shift towards an economic orientation in rural areas, such as the research sites, has resulted in changes to culture and livelihood, reflected in negotiations.

Increasingly, the decision to engage in collaboration or fulfil social responsibilities is based on individual assets, whether economic or social. Recently, farmers' economic assets have become limited due to the opening of markets and modern lifestyle, as narrated:

> I have 10 cong [13,000 sq m]. It is not enough to raise 3 children to eat and study, not all up to study in university. How much annual net income can one get from 10 cong?! Maximum VND 50 million [US$2,500]. One can get VND 3-3.5 million/cong for winter–spring rice, but spring–summer and autumn–winter [third season] are rather lucky or not; some seasons we can get just more than VND 1 million, but we can also lose in the 2 seasons, making maximum only VND 5 million per year in total. With the money, if the 3 children have big difference in age which means the oldest can start working and help the family, it will be okay, but with all studying in university and high school, it is difficult. In all, the monthly 4 million is finished with family expenses. Today, there are some costs that weren't there in the past. For instance, in the past, we used rainwater. Secondly, gas in the countryside; before, we only used the wood source from the garden. Thirdly, mobile phones; then about VND 500,000–600,000 per month for electricity. Those things already take more than VND 1 million. In the past, it only cost 2 litres of oil for lighting. (interview 22.11.2011)

Education, additional costs from replacing free local resources used in the past with increasingly available and convenient alternative sources amid labour shortages situation (e.g. machines), and other new needs for daily life have reduced households' ability to contribute to communal works. In addition, urbanisation, industrialisation and rice farming in the region have become more precarious (Mart A. Stewart and Coclanis 2011:112), the agriculture labour force has decreased (by 0.5 million people, or 2% annually) (Garschagen et al. 2011:106), and the migration within and from the Delta has increased, mainly to Ho Chi Minh City and Binh Duong (Huynh Truong Huy and Le Nguyen

Doan Khoi 2011:136). According to the Vietnam's 2004 migration census, 30% of residents in Ho Chi Minh City are migrants, and of these, 31.1% are from the Mekong Delta region (GSO 2004:125).

Subsequently, the out-migration and decrease of labour force create more obstacles to fulfilling the social requirements or established social norms for each individual. Consequently, co-operation behaviour, while still affected by social regulations to help and co-operate with neighbours, is increasingly influenced by households' financial and labour capacity. In some cases illustrated later in this analysis of field drainage, a lack of assets leads to hesitation to take on the risks of joining collective efforts. However, the challenges to collective actions can be significantly reduced by the maintenance of social cohesion, such as kinship ties. One informant states that:

> My family has farmed for 18 years. The pumping-out has been always together [with the other farmers]. I have 5 cong. When pumping, each household sets the pump and pumps on their own land. During pumping, there was no complaint about time or cost difference due to land-holding size. It is because all are brothers [relatives] in this zone. (interview 30.11.2011)

Although kinship has less influence in the Southern Delta communities than elsewhere in the country, there is also a popular adage: '*Near neighbours are more important than long-distance relatives*' [Vietnamese proverb]. Such social norms do add incentives encouraging individuals to act in the co-operative, responsible ways.

State planning for agricultural production and the sowing schedule

In contemporary Vietnam and the Mekong Delta, in particular, farming practices are primarily left to farmers to decide. However, rice cultivation is a matter which has an impact not only on farmers but also on national food security, which is a central state concern. Despite the increasing development of industry and service, strengthening agricultural productivity, husbandry and aquaculture remains a priority for the Mekong Delta ('Central Committee Strategy for Socio-Economic Development 2001 – 2010', CPV 2001:10-13, cited by Waibel 2010:11-12). In particular, '*losing the delta's rice surplus would be equivalent to the loss of national food security because the other regions in Vietnam barely produce any agricultural surplus*' (Fortier 2010, Wassmann et al. 2004, cited in Benedikter 2014b:95, Bucx et al. 2010: 21). A district cadre explains in more detail why production must be controlled by the state.

> If the farmers fail in the [winter–spring] season, it will influence the economy of the locality, then the overall economy of the city, and the country. It is because loss in only 1 season will lead to poverty for 2–3 seasons, which then affects the food security, failure to meet the rice export quota, economic impact … . If the

> farmer fail in the season, the local state can't collect fees and taxes; because the season is lost, it is impossible to collect money to construct rural roads, the fee for flood prevention, etc. Thus, it has vast consequences, so [the production] must be protected. (Interview 08.04.2012)

The state has a dominant interest in maintaining rice production, including yields and specific varieties of rice, to provide for national consumption and export. Consequently, the state monitors the process of field drainage through a sowing schedule attached to the overall production plan.

Every year, planning for the highly important winter–spring rice season consists of a production report assessing the agricultural production of the previous year and planning the current year's production. The planning report assesses general climatic and physical conditions which are used to decide the production targets for the province. The provincial plan is then translated into production plans for districts and communes. The planning report also introduces new models for rice, aquaculture and animal husbandry, on-going loan programmes for farming equipment, and solutions and guidelines for implementing the plan. The solutions include seed support, technical guideline and the training organisation for which the extension and agriculture offices are responsible. The timing or schedule for ensuring seasonal planning is translated into a sowing schedule. To facilitate the division of responsibilities and the implementation of the production plan, a standard procedure for transferring the plan to districts, communes and hamlets is set.

In sum, state concerns about production are translated into a designed structure for making plans and programmes that facilitate state management of agricultural production. The state structure requires the involvement of various state agencies, from management to technical agencies. The state plans and programmes for agriculture frequently are supported by scientific evidence, and implementation follows a rigid, standardised procedure for top-down transfer from province to district and then commune. Through this process, the Vietnamese state expresses its determination to apply infrastructure power through the state system of agencies, regulations and procedures to manage society. This process also shows the effect of modernism on the ideology of acquiring state legitimacy. The following sections explore these bureaucratic processes which reveal the state's characteristics.

Making of the sowing schedule

The sowing schedule at the centre of the practice of rice production is based on both farming traditions and scientific knowledge. DARD annually announces the sowing schedule before the start of the winter–spring season. The schedule indicates 2 or 3 periods of time during which farmers should start sowing. For example, the 2011 schedule had 2 periods:

1. Period 1: 17–23 November 2011 (22–28/10 lunar calendar[70])
2. Period 2: 16–22 December 2011 (22–28/11 lunar calendar)

The schedule is formulated seasonally by the Sub-Department of Plant Protection under DARD with 2 main purposes in mind. First, the schedule will guide farmers to start the season early enough to ensure a spring harvest and avoid damage from brown plant hoppers and related rice diseases. Second, the state encourages all farmers to start the season at the same time in order to facilitate mechanised farming[71] and contracted sales.

Technically, to formulate the schedule, the sub-department sets lamp traps for brown plant hoppers in various locations to measure their number and later test the check percentage of hoppers capable of carrying pathogens or viruses which cause rice grassy stunt and rice ragged stunt diseases. The sub-department tracks the wind direction to simulate the migration pattern of the hoppers, conducts field observation to examine the situation of rice grassy stunt and rice ragged stunt diseases, collects data from the Southern Plant Protection Centre and other provinces and, last but not least, checks the water level (interview 06.02.2011, Figure 5-11). The winter–spring schedule has been issued annually since the early 2000s, and the plant hopper factor has been added since a 2006 epidemic of the insect and rice grassy stunt and rice ragged stunt disease. The schedule serves a slogan for agriculture production planning: '*Sowing collectively, at the same time, and avoiding brown plant hoppers*' (Sub-Department of Plant Protection Can Tho city 2011).

[70] Cultivation is traditionally related to water, tidal and culture, so the lunar calendar is used in all events in farming practice, includoing the state schedule.
[71] Contracted pumping and mechanised harvesting are possible in large areas, more than 40 ha and more than 10 ha, respectively (interview 16.11.2011).

Figure 5-11: Procedure for setting the state-sowing schedule in Can Tho (based on 2011–2012 field study)

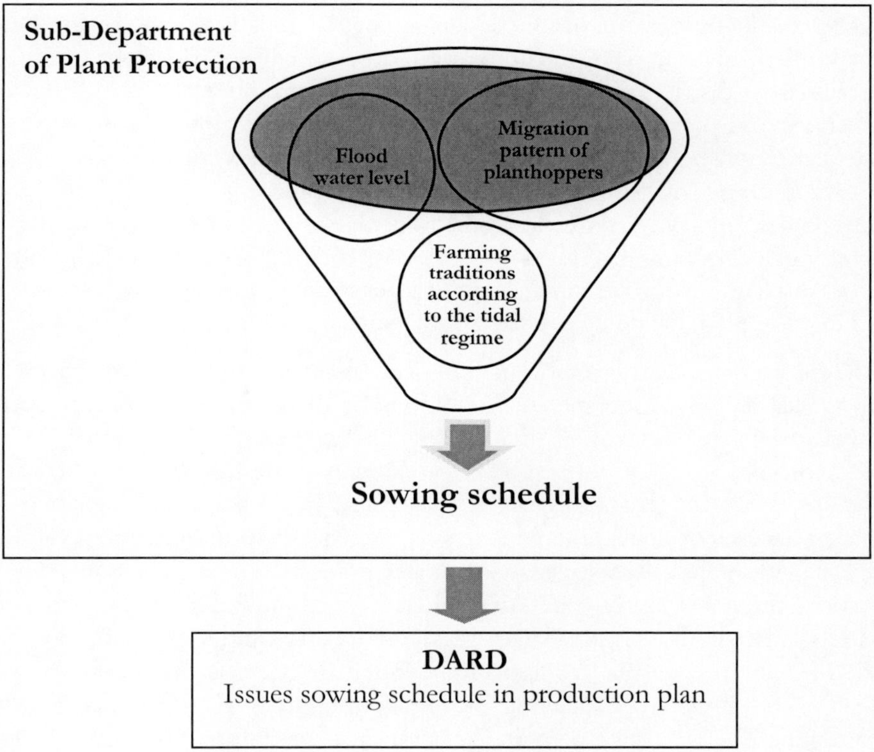

Managing sowing and field drainage from the modern state perspective

Field drainage, with its link to important agriculture production, is subjected to the influence of the modern state. As in other sectors, the management of the activity follows standardised planning and implementation.

> The practice of top-down state management and collectively organised production has by no means vanished from the state agenda, despite the return to household-based production. (Benedikter and Waibel 2013:23).

To ensure food security and maintain the state's status, state agencies and regulations continue in production management, specifically through a standardised bureaucratic procedure with meetings and mandate transfer from the province to the commune.

Rigid bureaucratic procedure for sowing schedule dissemination

After being formed at the provincial level, the sowing schedule, along with other considerations for production planning, is transferred to the local state for implementation. The transfer is done in a strictly top-down process from province to district, district to commune, and commune to hamlet through meetings and official documents (Figure 5-12). Despite the uniform structure for plan implementation, significant and (informally) accepted deviations arise in what come out from the meetings and practices. What allows these differences, and why? After a production plan is formulated at the province level, DARD organises a meeting with representatives from the DARD offices of all districts. At the meeting, district cadres receive mandates to carry out the production plan, including the sowing schedule.

After drafting a district's production plan based on the mandates from the provincial level and the specific conditions of the district, the district DARD office calls a meeting of district cadres (district People's Committee leaders, plant protection office cadres, irrigation station cadres), communes' irrigation and rural road cadres and district extensionists. This second meeting serves the same purposes of delivering mandates to the local level and disseminating state policy without much discussion of actual practice. The production plan for winter–spring season is then drafted at the commune level by either the extensionist or the irrigation and rural road cadre, under the supervision of CPC leaders responsible for agriculture (often the vice president of the CPC). This is followed by a meeting at the commune level. Until reaching the commune level, the state planning structure for agricultural production follows a standardised pattern of hierarchical system in which policy and decisions are made at higher levels and transferred from the top down to the local level for implementation.

Figure 5-12: Ideal transfer of the sowing schedule (made by author)

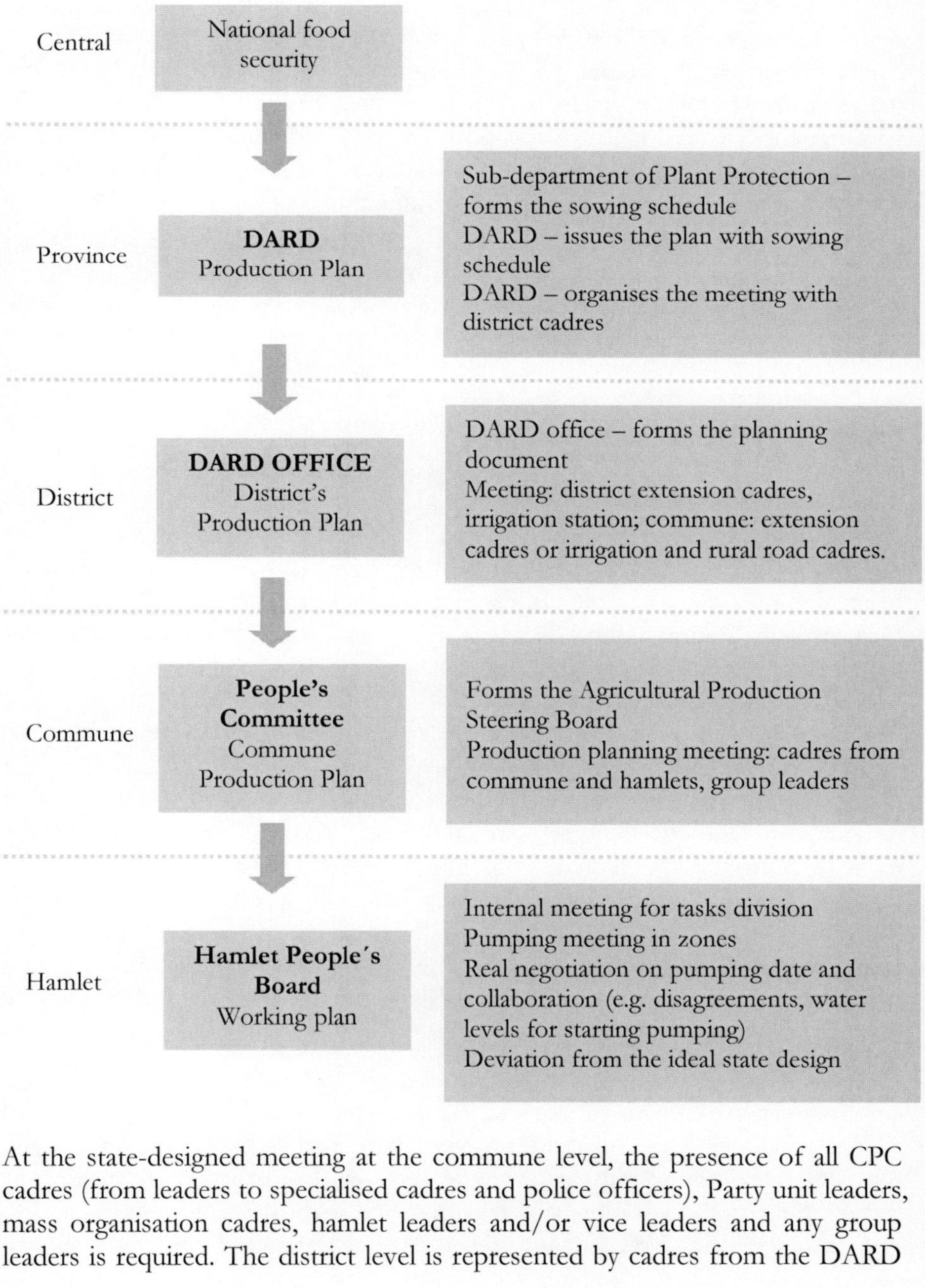

At the state-designed meeting at the commune level, the presence of all CPC cadres (from leaders to specialised cadres and police officers), Party unit leaders, mass organisation cadres, hamlet leaders and/or vice leaders and any group leaders is required. The district level is represented by cadres from the DARD and extension offices. The meeting observed was officially called the *Conference for Reviewing the Agriculture Sector and Implementing the Production Plan for the Winter–Spring Season in 2011–2012*. As indicated in this name, such meetings consist of 2 main sessions. First, the production situation in the commune from previous seasons is reviewed. Second, the CPC representative informs attendees about the plan for agriculture production in the winter–spring season, including the sowing schedule. As well, a resolution list for Production Steering Board is announced, with the involvement of various cadres at the commune and hamlet levels (Figure 4-6, Chapter 4). The board's function is described as follows:

> The winter–spring production steering board year 2011–2012 has the responsibility to organise propaganda to the broader population, enhance the people's awareness about the brown plant hoppers, rice grassy stunt and rice ragged stunt diseases, prepare the field before sowing the winter–spring rice, not using the plant hopper-susceptible varieties, sow together at the same time according to sowing schedule. (Decision of Commune People's Committee 199/QĐ-UBND — 28.10.2011 — To form the Production Steering Board in Truong Xuan B for the winter–spring season 2011–2012)

The meeting is directed by the CPC vice president responsible for the economy and facilitated by the irrigation and rural road and/or extensionist cadre. The meeting also serves as lesson-learning event in which management weaknesses from the previous year are criticised. Although the meeting is somewhat formalised without much discussion of the coming practice, the designed series of meetings serves as the annual rehearsal of administrative structure in Vietnam. As well, agricultural policies today can be stated as only encouragement, so the meetings serve as a tool to raise awareness of the importance of production and state management. Through the meetings, the local cadres are deployed for implementation, which primarily takes the form of mobilisation. Without all these steps, the superior state agencies fear that the requested production targets will not be met. However, the meetings themselves are only the start of complex processes to implement the sowing schedule. Indeed, the meetings can be seen as an incomplete process of negotiating and discussing practice, which must be followed by later negotiation among state cadres, farmers and sometimes private actors at the interface of the field drainage process.

Looking closely at the commune meeting, it resembles a formal event at which commune cadres inform hamlet and group leaders of the production plan and sowing schedule. While discussion about the schedule takes place, the rigid, formalised procedure and the mandate received from provincial level limits district and commune cadres' ability to respond with official accommodations. For instance, the 2011 sowing schedule was strange based on farmers' experience as it had a long, 1-month interval between the 2 sowing periods. Normally, the interval is one 'water cycle', which is around half a month, so that in a month-long period of October or November (lunar calendar), the 2 periods of low water can both become sowing times. The issue was raised in all meetings observed in the Thoi Lai district. A scientific reason was given as explanation. According to district cadres, the provincial agency had researched the wind direction and migration pattern of the plant hoppers and concluded that the middle low-water period (30.11.2011–06.12.2011 or 06.11–12.11.2011 in the lunar calendar) would be susceptible to a brown plant hopper epidemic[72].

Further discussion of the schedule did not happen in the meeting, and no agreement could or was necessary to be reached for several reasons. First, neither district officials nor state cadres at the commune or hamlet level have the authority to amend the schedule. Second, regardless of what is discussed during the meeting, there is a co-understanding that the sowing schedule, or agricultural policy in general, takes the form of encouragement. Although the local cadres mobilise farmers to implement policy, farmers who decide what farmer practices to carry out at a selected time. Thus, at the commune's production planning meeting, expectations for achieving consensus or an ideal plan are low. Instead, the goal of bureaucratic meetings is dissemination.

Regarding commune cadres, their presentation of the provincial sowing schedule at the commune meeting does not equate to a lack of local knowledge. Neither does it mean that the Vietnamese state's top-down control of policy implementation leaves no space for local negotiation and adaptation. Rather, it demonstrates the local cadres' choice of arrangements to best suit diverse conditions: the state interest in control through a standardised, bureaucratic structure and local arrangements adapted to particular natural, physical and social conditions. However, this process of choosing the arrangements is done

---

[72] Brown plant hoppers have a lifecycle of almost 1 month divided into 3 periods: egg, larva and adult. Only adult hoppers can migrate between fields. Thus, the adult period of 7–10 days old is the most dangerous period for young seedlings and needs to be avoided. The plant hopper avoidance strategy (*né rầy*) is carried out as followed: When the rate of hoppers coming into lamp trap reaches the peak number, farmers start seeding, soaking and sowing 2–3 days later. The young rice seedlings then come immediately after the peak period of adult hoppers, minimising risk. The risks coming from other stages in the hopper lifecycle in between are addressed by covering rice plants with water (*om nước*) as brown hoppers favour of dry, sunny conditions (Le Hoang Vu and Thach Thao, 2013) http://nongnghiep.vn/nongnghiepvn/72/2/2/106817/Ne-ray-om-nuoc-hieu-qua-SX-lua.aspx, accessed 27.05.2014.

in the later state of the practice. At this stage, the commune meeting and production plan serve to fulfil the state's bureaucratic mandate. The incomplete discussion and agreements are supplemented by subsequent negotiation and flexibility in practice. Although the existing bureaucratic checklist of tasks and meetings remains to be rigid with little flexibility in the outcomes, conflicting natural, physical and social conditions make the implementation of state plan difficult to predict or judge.

Until the commune meeting, the process of relaying the sowing schedule serves a bureaucratic goal to disseminate state policy to lower levels which does not necessarily correspond to reality on the ground. This process illustrates the Vietnamese state's form of mobilisational authoritarian or state corporatism, in which the state, through its structure of agencies and regulations, or infrastructure power, wishes to '*mobilize people to support its programmes and policies, maintain channels of communication between authorities and each sector of society, and manage social and economic groups that otherwise might become unruly*' (Kerkvliet 2001a:3). Adding to the structure, the sowing schedule has been adapted to the local tradition of sowing in October and the scientific knowledge of sowing at the low water cycles so that the application of state planning is identical with farmer practice. It takes time for the feedbacks to be adapted because an effective, quick feedback pathway is not in place. Thus, based on this level of state responsiveness, the state can be assessed as indirectly or incompletely responsive.

Still rigid bureaucracy but starting negotiation at the hamlet level

Receiving the mandates from the commune state, hamlet People's Board calls for an internal meeting to inform all hamlet cadres about the sowing schedule. In the meeting, cadres are given responsibilities for organising and monitoring pumping meetings, pumping and sowing for zones. This meeting is followed by the pumping meeting at different zones, with the attendance by all farmers in the zone and led by hamlet cadre(s) and any group leaders (if there is).

> The internal meeting of the People's Board is necessary because it is not the case that all hamlet cadres join the meeting at the commune. Therefore, we need to meet for everyone to be informed and discuss the plan, so that when farmers ask, we all know how to respond and for group leaders to know in order to organise in their groups. (Interview 23.11.2011)

The statement that the purpose of the hamlet meeting is '*for everyone to be informed and discuss the plan, so that when farmers ask, we all know how to respond*' demonstrates that the meeting is a step in the bureaucratic checklist that local state officials (state assistants) have to fulfil, although the nature, physical and social aspects are often divergent, making practice difficult to predict. The meeting also usefully creates buy-in and demonstrates state involvement in local affairs. Ultimately, in addition to showing responsiveness to diverse, local natural, physical and social conditions, the hamlet meeting remains part of the Vietnamese state's attempt to simplify and standardise society through the administrative ordering of nature and society (for a discussion of the nature of a state, see Scott 1998:2, 4). The state is always concerns about whether a policy is passed on to all parties involved. In this context, the purpose is to maximise the possibility that each local cadre and farmer will carry out the desired cultivation. Thus, when the first hamlet meeting did not have the participation of everyone requested, the superior state officials felt obliged to compel people to meet again in another meeting or through other platform.

Although the hamlet meeting is part of a rigid bureaucratic checklist, it also enables the state structure to get in touch with ground-level reality when hamlet cadres take local conditions into consideration in production planning. Cadres at the hamlet level are community insiders, most of whom are respected farmers, having demonstrated not only for good farming practice but also high social spirit in recent years. It is stated that '*From the* [internal hamlet] *meeting, the season gets started*' (interview 11.11.2011). Now starts the planning for field drainage in all areas of the hamlet with or without the presence of a farmer group, with the facilitation of hamlet cadres throughout the process from pumping meeting to sowing.

In the hamlet, all or several members of the hamlet People's Board are responsible for monitoring field drainage and sowing in certain zones. Field drainage is one sub-task in the contested process of disseminating the sowing schedule. The appointed board members and any group leaders (e.g. in Dong Thuan commune) organise and facilitate the negotiation in the meeting. While responsibility is individual, overall management is mandatory; thus, 1 hamlet leader (either the hamlet leader and vice leader or the secretary of hamlet Party Unit) has to be present at the pumping meeting.

Conclusion: Adaptation of the rigid bureaucratic structure

In short, state policy is transferred through a rigid structure following a standard procedure in order to fulfil bureaucratic goals which do not allow much response from the grassroots level. Through this stage, some would assert that this model reveals the dominating or authoritarian mode of the Vietnamese state, with top-down decision making, dissemination without consensus, a technocratic orientation, a checklist, and lack of local participation. However, policy implementation at the lowest level, that of practice, reveals another picture. While the meetings at the district and commune levels mostly serve the purpose of information dissemination, the internal meetings at the hamlet level and the pumping meetings with zones and group leaders feature more practice-oriented discussion, leading to agreement or contestation. The local negotiation and flexibility ensure a smaller risk of failure from '*in the sky*'[73] policies. One might question to what extent and at which stage in the state's procedure, negotiation, including that over local conditions, should be involved in order to increase policy effectiveness. To better capture the options and possible results, an ethnographic research on the possibility and extent of local participation on decision making is necessary. In the present system, adaptation in the form of trial and error takes a long time when policy makers receive official information about failure and changes only at the end of the season. Consequently, any considered changes can be applied only in the next season (year).

The implementation of top-down policy at the local level reveals a structure of de-concentrated management and rigidly top-down bureaucratic procedures involved in disseminating policy. Above all, the state wants to be, and is, present in every aspect of farming production and rural livelihoods, claiming a leading and facilitating role in the process, including in field drainage. The question is whether rice farmers cannot start their individual work before the state plan because of dependency on state support or an obligation to follow the plan of the controlling state. Neither explanation is applicable to this case. Farmers are the masters of the land. With significant empirical experiences, they become experts on their own land and generally capable of farming individually or in co-operation with their neighbours. However, the rural Mekong Delta has never been a free zone. Especially since the 1975 reunification and since the 1990s when the Delta's agricultural production become increasingly important in supplying food and income for the whole nation, the state has maintained a presence in the region. State agencies are involved in every step of the farming process down to the farming level. With the intensified irrigated agriculture and presence or penetration of the state, farmers' production and livelihood connect to state policies and rely on state support at certain points. In the Delta today, there is a system of interdependency between cadres in the office and farmers in the field.

The growth of this interdependency derives from the long history of interaction and negotiation between the policy-level state and the local level. After years of

---

[73] Commune cadres used this term to describe policy which is so much far away from reality that it is impossible to implement.

(trying to) coordinating agricultural production in the Mekong Delta, the commune, district and provincial state recognised the established farming practices and conditions at the local level, including sowing as close as possible to early October and sowing in the low water period. Learning from the failure of the command-and-control system, contemporary state agencies accept and incorporate the diversity of local practices into the production plan. These agencies adjust seasonal planning to occur at farmers' desired timing — at low water points and with the first period always falling in October — and either officially or informally allow space for adjustment of implementation to local conditions. Therefore, the timing of commune production meetings and hamlet pumping meetings is arranged depending on local natural and physical conditions, mostly the water level and dyke infrastructure.

In turn, farmers and local cadres also make adjustments to follow the state plan. However, instead of simple compliance, their acceptance of policies and strategies is selective and given only when the policy gains farmers' trust through a pilot project or proven success. Regarding the sowing schedule, a state cadre states:

> In the first year, it was generally new to people. Some followed the schedule, but some didn't. The ones who didn't follow the schedule, some were later affected by the brown hoppers, so then, they slowly realised the suitability of the schedule. Today, the percentage of the farmers following the schedule is getting higher and higher. (interview 06.02.2011)

The limited attention the sowing schedule receives from the farmers was achieved only after the 2006 brown hoppers epidemic when, across 3 seasons, approximately 448,305 ha (10.66% of the total sowing area of the Mekong Delta) were infected by brown plant hoppers, and 175,764 ha were damaged by rice grassy stunt and rice ragged stunt diseases. Together, these caused thousands of VND in harvest losses and costs for the elimination of hoppers and destruction of infected plants (Department of Plant Protection, 2007, cited in Pham Van Lam...). The clear threat from the hoppers gave the state an opportunity to legitimate its policy and role. The state strategically took advantage of that opportunity and added hopper avoidance to the sowing schedule. Subsequently, the state gradually gained respect from Delta famers.

In the contemporary Mekong Delta, farmers accept the state's suggestions only once their legitimacy is proven. As seen in recent events, the legitimacy of the power of scientific-based modernism faces challenges. Recently, in the Delta, the state and scientists have had to demonstrate their capabilities through technological suggestions. First, by ceasing to impose coercive power (Göbel 2011) since Doi Moi, the state has eliminated the possibility that it can force farmers to comply with state suggestions. Second, the farmers are less likely to accept certain scientific suggestions as many local arrangements work well. Among farmers for whom rice cultivation is the financial backbone of their livelihood, the inertia of established practices and the fear of risks associated with new suggested techniques are the main challenges for scientific and management agencies implementing any new technology.

To implement policies, the state has applied discursive power, along with the infrastructure power expressed in state system of legal regulations and structure of agencies from the central to the hamlet level. Evidence of discursive power arises from the decade of intensive propaganda about the sowing schedule disseminated by state mobilisation bodies, including extension and agricultural offices, CPCs and commune Farmer's Unions. With the continuous mobilisation efforts in the dense structure of agencies and regulations and strategic policy implementation (do it in case of crisis), most farmers understand and seriously consider the sowing schedule when selecting farming practices. This wide acceptance by farmers was seen at all observed hamlet pumping meetings, as elaborated in the next section on field drainage from farmer perspective).

Thus, the state continues to attempt to exercise discursive power through mobilisation strategies. However, the state has not achieved its planned result of making farmers think in the way it wants them to. For instance, in the case of the sowing plan, the decisive water level and other factors, along with the state schedule, are taken into consideration to determine the sowing date and so the pumping date. Therefore, in exerting discursive power, the state agencies found themselves caught in a negotiation interface in which not only the state but also the farmers have the power to influence the agenda. The process for implementing the state sowing schedule, which is analysed thoroughly in this chapter, does indeed create a dialogue between the state and the involved stakeholders, demonstrating the dialogue facet of the Vietnamese state (Kerkvliet 2001b, Kerkvliet 2003).

This analysis of the state's formal structure to transfer the sowing schedule has answered the question of how the state envisages its role in achieving control and management over production in the Delta, including field drainage. How the state plan is implemented and transformed into practice is analysed hereafter, including all the processes for organising, monitoring field drainage process and deciding the sowing schedule at the commune and hamlet levels. Although all matters surrounding the sowing schedule have been adapted to science-based state policies and local farming practice, mutual agreement on areas and seasons is still far from complete. Instead, the implementation of the sowing schedule and farming practice is seasonal and local. It goes through a rehearsal of negotiation at the interface (the field drainage in communes and hamlets) where local cadres and farmers consider factors, including the state agenda and natural, physical and social conditions. In general, the field drainage process is lengthy as decisions on collaboration, interaction and negotiation are built on individual and communal consideration, economic and social benefits, state policy and local farming ideologies. The process plays out step by step in field drainage, from agreeing on the pumping schedule to meetings before pumping, during and after pumping (Figure 5-10). The interdependency and negotiation of interests by various stakeholders through farming activities are also explained.

Sowing schedule and field drainage from the ecological and farmer viewpoints

Picture 5-5: Gathering before the start of pumping, Thoi Lai district (Huynh Linh, 12.2011).

Wandering around the hamlets in September and October (lunar calendar), one can easily witness various ways farmers interact with local cadres about field drainage and sowing-related topics. The discussions are mainly about when to start pumping, when to sow and how to do it, whether alone, with neighbours or through contracts with high-capacity pump owners. Indeed, farmers have proactively planned their farming practices in conjunction with the state planning process. Before the season starts, farmers prepare and plan for the seed supply, sowing and all cultivation practices for their own rice field. The co-existence of the 2 structures — the state production planning and farmers' or local arrangement for production — results in a complex process of interaction and negotiation with agreements and contradictions.

In the researched context, the implementation of the sowing schedule consists of a process of interaction and negotiation among local state cadres, farmers and farmer groups to figure out the most suitable timing and practices. The negotiation process might have more agreement, contestation, conflict or conflict resolution. During this process, local farming practices, scientific knowledge and state policy are rehearsed and re-tested. Above all, the state has learned the necessity of making directives, such as the sowing schedule, flexible. Various conditions, especially the water level, dyke conditions and farmers' choices, strongly influence sowing practice. However, complete freedom in local practice is counter to the state's desire to control the output of the Delta, in which it has an interest. In contemporary top-down Vietnam, the state wants some control and monitoring power down to the grassroots level (i.e. directives such as sowing schedules), but the Communist state of Vietnam today is different than the rigid, one-size-fit-all policy of the 1970s–1980s collectivisation period in the.

While farmers' practices and local arrangements influence state policy implementation and sometimes even policy itself, they, too, are adjusted to accommodate the state bureaucratic procedure for production. Thus, many local practices at the commune level follow the general pattern or timeframe planned by the state (see Figure 5-13). As mentioned, after the commune production meeting, an internal hamlet meeting is held, and the winter–spring season starts. The hamlet meeting, on one hand, is the end of the state bureaucratic checklist; on the other hand, this meeting initiates the organisation of field drainage. The field drainage process, mainly done by collective pumping, has 3 stages: preparation (pump preparation, dyke repairs and pump setting), pumping and cost sharing. All these stages have the involvement of various stakeholders and are influenced by natural and physical conditions, social institutions and state incentives carried out through policies and mandates.

Figure 5-13: Process of organising field drainage before sowing for the winter–spring season

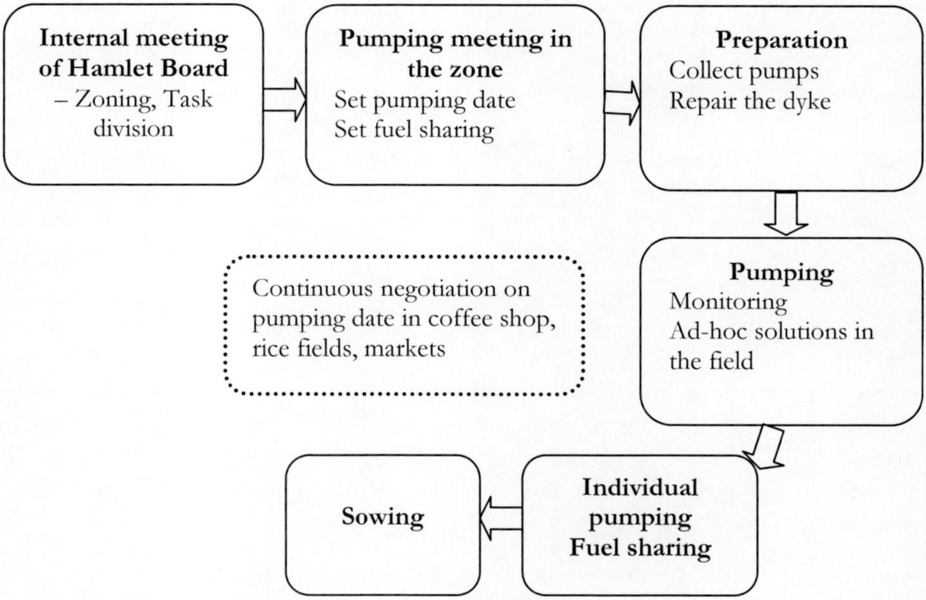

In addition to securing consensus amongst farmers about when to start pumping, closing and repairing all sections of the covered dyke is essential during the preparation stage. After a long period of inundation from natural flooding and deliberate usage of flood water to collect nutrients for the field, many dyke sections are weakened or even broken. Therefore, to practice field drainage in collaboration with neighbouring farmers, all households have to agree to check and repair individual dyke sections that help strengthen the whole covering dyke. All involved should collectively repair the largest dyke sections. When pumping starts, monitoring pumping, responding to any crises (e.g. dyke damage, conflicts) and sharing fuel are the main activities. After pumping, fuel or money compensation continues in order to complete any necessary sharing even as preparation for sowing begins.

All steps of the process are vital to the success of pumping, which allows on-time sowing and subsequently contributes to the success of the important winter–spring season. The cost-sharing arrangements affect individual satisfaction with social co-operation, which influences present and future social cohesion and incentives to join communal work. The cost-sharing problems especially affect zone or group leaders who are not hamlet cadres but spend time, labour and money providing for the community welfare. Given the importance of all stages of the process, the state monitors each step through the network of hamlet cadres, who periodically report on hamlet production to the responsible commune cadres, including the irrigation, extensionist and farmer's union cadre.

Agreeing on the pumping schedule: The meeting

Agreeing to or at least accepting a nominal pumping schedule is part of a long process, in which hierarchy, technocratic ideas, local farming traditions and negotiation are overlaid upon various meetings and forums. This section analyses the chronological order of the process of making a pumping agreement.

Zoning: Farmers' consideration of state ideology

Zoning, or dividing farmers into groups for collective pumping, is organised by hamlet cadres with farmers' input. The present-day arrangement for pumping zones is the result of interactions, including negotiation and adaptation, between the policy-level state and local stakeholders, including state cadres and assistants, farmers and others.

Farmers decide zones based on practical criteria, such as water level, the boundaries of the existing canal system and the condition of covered dykes. Farmers whose fields are separated from others by low ridges (which are still inundated at the end of the flooding season) and share a good enough (August) dyke make up a zone (Picture 4-2 Chapter 4, Picture 5-6). State cadres often identified the zone as group. Zone boundaries overlap with those of either an old co-operative group (*tập đoàn sản xuất*) or a registered production group. Otherwise, new zones are designed for a specific season according to water level, dyke condition and farmers' consent. The co-existence and, frequently, the overlap between pumping groups (state ideology) and pumping zones (farmer practice) illustrates the similarity in local stakeholders' and the state's interests. Both aim to secure production, and by promoting groups, the state fulfils its task of encouraging a collective economy, which is one of the critical principles of a communist regime.

Picture 5-6: Top – inundated field with covered dyke (where trees are); Bottom – the dyke (farmers are choosing trees for dyke repairs, Huynh Linh, 2011)

Today, agreeing to working in zones is a result of farmers' reconsideration of collective farming since the collectivisation failure in the 1980s. With the expansion of irrigated agriculture to facilitate double and triple rice cultivation, collective field drainage in zones has become a practice for farmers to deal with common threats. In this way, farmers and local cadres rehearse the state policy of collective farming adapted for intensive, commercialised agriculture production. This rehearsal of old institutions is possible only due to the state's adjustment of policy (introducing a new form of collective action) and persistent mobilisation or application of its infrastructure and discursive power through the network of state agencies and regulations.

Continuous infrastructure investment, domestic usage and farming culture significantly decide the zoning practice. Zoning is based on the hydrological boundaries of canals and dykes built for practical and economic reasons and thus relies on the existing infrastructure built over the history of human intervention in nature. In addition, zone boundaries often fall within the administrative boundaries of the hamlet or commune due to the overlap between administrative and hydraulic boundaries. In the process of imposing standardisation and simplification in order to manage and penetrate all of society, the state has adapted to cultural and farming traditions. Zone boundaries are subject to change as field drainage is highly dependent on climatic conditions which fluctuate over time. Zones also rely on farmers' commitment to participate, from agreeing on the date for collective pumping to repairing and closing the opened dyke sections to fuel sharing.

In short, zones are both a convenient bureaucratic and a systematic way to connect people in a forum for social co-operation and state involvement. Also, given the domination of practical rationales in forming and operating zones which sometimes coincide with administrative boundaries, zones reflect the economic interdependence which has played an important role in social arrangements of the Mekong Delta.

Pumping meetings in zone: Formal efforts to guide practice

After setting zones based on predicted and past conditions and boundaries in the previous season, either local cadres or farmers are assigned as the leaders of each zone. These leaders are in charge of identifying possible pumping dates according to the water level, dyke conditions and state sowing schedule. Pumping meetings combine the formality of following the bureaucratic procedure applied at higher levels and the informality of social arrangements for collective field drainage.

When a pumping period is identified, the zone leader, along with the leaders of hamlet People's Board, calls for a meeting — namely, the pumping meeting — at least 1 week before the planned pumping date. The 1-week interval is necessary to give time for dyke repairs and pump preparation. The pumping meeting takes place in the hamlet Information House, usually the official office of the hamlet; the house of a cadre or a prestigious farmer whose land is in the zone; or a public place, such as cafeteria (Picture 5-7). Different atmospheres were observed in these different sites, more official and rigid at the Information House and more relaxed in individual houses or public places. However, the difference only happened at the beginning of the meetings; the later discussion and negotiation in all the observed meetings exhibited a similar tendency for free talk and equal negotiating power (further analysed in this section).

All farmers who have land in the zone, including *xam canh* farmers (those living in other hamlets), are invited to the pumping meeting. The zone leader informally gives the invitation in person by walking around the hamlet or by phone. The tradition of informality at rural community has been challenged recently by farmers' busier, more mobile lifestyle. Farmers today engage in more income activities than rice farming. Consequently, some experienced hamlet cadres' habit of extending a morning invitation for an afternoon meeting that same day no longer works and has been criticised for failing to gather farmers for the pumping meeting (observations in the 2011–2012 winter–spring season). Existing institutions are under pressure to prove the suitability or legitimacy of current arrangements amid on-going socio-economic changes.

Getting all farmers involved to attend the first meeting is impossible. In many cases, only few farmers come to the first meeting, so several meetings are needed to achieve consensus on dyke repairs and the pumping schedule. The low participation rate results from farmers' involvement in other activities or the length of time until the critical moment for field drainage. In the 2011–2012 winter–spring season, high water levels and long-time retention of flood water were the main reasons for low participation in early meeting. Consequently, in this pumping season, many zone leaders and local cadres had to re-invite participants 2 or 3 times until negotiation was possible. In some cases, the third meeting was not necessary because the water level (November 2011, lunar calendar) had fallen lower than the dyke, enabling gravity drainage.

Picture 5-7: Pumping meetings in zones:

Top-down: Meeting 24.11.2011 (Dong Thuan commune) at a hamlet's Information House; Meeting 29.11.2011 (Xuan Thang commune) at the house of a prestigious farmer who was later nominated as zone leader; Meeting 12.11.2011 (Truong Xuan B Commune) at the house of a hamlet cadre; Meeting 29.11.2011 (Xuan Thang Commune) at the house of a prestigious farmer who used to be a cadre of the hamlet's Farmer's Union and leader of the gardening club in the 1990s and is the zone leader for field drainage.

* A hamlet's Information House serves as a meeting place for such occasions as pumping meetings, meetings to transfer state policies and organising elections. The house acts as the bureaucratic symbol of the state structure in the hamlet.

The incentives for pumping meetings vary among zones and farmers in the same zone. Localities have diverse water levels and dyke systems, while farming households have different pump availability, labour resources and individual sentiments towards communal work. Alongside the diversity that significantly affects the process of reaching agreement, an informal community network generally complements the state organisation of field drainage. While it is important to hold meetings to reach an agreement, the pumping meeting, in many cases, becomes redundant to informal community network but nevertheless is considered a necessary part of the bureaucratic decision-making procedure for field drainage. The diversity in farmers' incentives and social arrangements makes the effectiveness of state mobilisation only one piece of the puzzle contributing to the final result of the field drainage and sowing activities. Although the pumping meeting cannot provide the final decision for field drainage, it creates an interface where hamlet cadres, group leaders and farmers enter a process of negotiation and adaptation.

Choice between formality and informality in state promotion of grassroots democracy

The pumping meeting can be facilitated in 2 ways. The formal method includes several standardised steps: introducing the delegates and purpose of the meeting and announcing the commune meeting's results, including the sowing schedule[74], followed by discussion. In contrast, the more direct, informal meetings have only a very short announcement of the state schedule, immediately followed by discussion. While both styles fulfil the state mandate to disseminate the sowing schedule, the differences are derived from individual methods of mobilisation which are based on experience.

Experience gives local cadres more power and know-how to experiment with the state's mandate. Indeed, young, newly elected cadres tend to make the meetings more formal, while experienced cadres (some have even held the position since reunification in the 1970s) favour informal, direct interactions in the meetings they facilitate. Consequently, the way the sowing schedule is announced differs among hamlets. While in some cases the schedule is described as an obligation allowing farmers only to choose which the best date

[74] Note that the presence of the researcher also increased the degree of the formality of the meeting.

in the state schedule, other hamlet cadres state '*the state schedule is like this, but you and I* [the cadre and farmers as the hamlet leader is both group leader and farmer] *shall see the water and decide when we can do it* [the sowing]' (meeting 24.11.2011). The latter way which references farmer's planning proves the co-understanding and adaptation of the cadres on the ground, who consider such prominent factors as natural and physical conditions, water level and dyke system. This consideration reveals the continuity of deviation from the central plan at the local level. Local deviation was especially observed in Southern Vietnam after 1975, highlighted by the fence-breaking from collectivisation in the 1970s when local cadres and politicians experimented with market forces (Porter 1993, cited in Benedikter 2014a:33–34).

However, the choice to announce the sowing schedule as an obligation does not necessarily indicate a lack of local knowledge. Rather, the action shows the pressure that a mandate from the CPC exerts on a local state cadre. Young and new hamlet cadres most frequently act in this manner. While proving to be loyal, these cadres have not yet learned to apply flexibility and adapt state mandates to local conditions as established cadres have.

Hamlets' flexibility occurs with the acceptance of the commune state.

> The sowing schedule given from the higher level, according to the grassroots democracy regulations, we [local state] have to let the inhabitants informed and discuss. According to the general regulation, we have to encourage the people to follow the schedule in order to avoid damage. However, when something bad happens, for instance, in this year when water level was still so high, the commune [state] reported to the higher level, and the superior state sent someone to reassess [the situation]. If it is the case, the support later on [the state support for the households that experience damage during farming production, and as a form of encouragement, regulations limit state support to households that follow the state sowing schedule] will not only be limited to the households that sow within the state schedule. (interview 30.11.2011)

The choice to engage in formality and the tendency to promote the image of the democratic state demonstrates state learning and adaptation. The state adjusts its management to permit flexibility, as well as informality, at the hamlet level. However, the state does so strategically in keeping certain roles in the local affairs. At this local level, state involvement is performed by local cadres who are also part of local affairs as farmers and neighbours. Closed to and understanding practice and reality, local cadres adjust the state schedule to be realistic and unofficially give value to local, established arrangements and institutions. However, whether the ensuing negotiation provides a successful outcome depends on the parties involved.

In addition, a prominent aspect of meetings is the promotion of grassroots democracy which, taking a broad view, characterises the Vietnamese style of democracy: freedom with mobilisation.

> We, the local state, only give the idea. You [the farmers] are the ones who discuss and decide. (pumping meeting, 23.11.2011)

This idea was repeatedly promoted in all 17 observed pumping meetings to affirm that pumping meeting is a democratic event and that the state is only there, upon request, to support farmers. The hamlet cadres constantly confirmed the state's image of letting farmers decide their own business of farming, which the state facilitates and supports. This action is part of the mandate to implement Grassroots Democracy in Vietnam according to Decree - 79/2003/NĐ-CP-, issued in 2003 and replaced by the Ordinance on Implementing the Democracy in Communes, Wards and Towns - 34/2007/PL-UBTVQH11. This ordinance states:

> The people directly discuss and directly decide the policy and the contribution rate to construct infrastructure, to form the social welfare within the boundary of commune, village, or people zone which is fully or party funded by the people; and other internal works of the community which are within the legislation. (Article 10, 34/2007/PL-UBTVQH11)

According to the ordinance, residents may decide many internal community affairs.

The enactment of the legal set of grassroots democracy has gained the attention of developers, donors and academic scholars. However, the democracy promoted in Vietnam is far from similar to 'Western-style democracy. According to Kerkvliet et al. (2003) and Minh (2008),

> The promulgation of the GDD [Grassroots Democracy Decree] was primarily an acute response by Party and government to large scale rural unrest during the 1990s, including demonstrations and violence against local officials [in order to] demonstrate decisive actions on mismatches between espoused commitments to 'good governance' and actual practice. (cited in Larsen 2011:318).

Adding to the relaxation of social tensions, the Grassroots Democracy Decree increased or re-established state legitimacy (ibid) and prevented increasing challenges to the Party's control (ibid:325). Although there is no concrete evidence of the value of the legal reforms, the Party and state strategically and successfully passed the challenge of overt resistance at the local level in the 1990s. This survival can be seen as the success of Vietnamese Leninism in both overcoming the loss of a centrally planned economy and generating donor supports in the early 1990s, thus benefiting citizens (Fforde 2008a:13). A similar analysis can be applied to China. The Chinese Communist Party decided to introduce and experiment with democratic elements in order to stay relevant to the needs of a more complex, diversified society (Zheng and Lye 2011:40). In that sense, the top-down reform in China does not mark the Communist Party's retreat from controlling the countryside but, rather, a strategy to strengthen the Party's rule by accommodating rural democracy (ibid:33).

Although Vietnam has been influenced by a market orientation and the international agenda of donors since the 1990s, the Party and state's responses and adaptation to local requests do not show any signs of donor push but, instead, internal decisions made by the Party and state (Larsen 2011:325). This decision-making process signals the continuity of rulers' nature recorded with the history of peasant rebellions under the centralised authority of the Confucian monarchy or colonial powers (Woodside 1970, 1989, Marr 2004, cited in Larsen 2011:328). Thus, the case of Vietnam in the 1990s raises the question of whether the status of the Vietnamese party–state system as an authoritarian regime should be rethought. In an analysis of language, local arrangements in leadership election, and local officials' acceptance of informal farmer groups, Adam Fforde argues for this viewpoint.

> This may be interpreted in a variety of ways, of which one of the most interesting is the idea that Vietnamese Communism had either long become accustomed to such gaps between its theory and reality, or that it simply lacked the political power and will to take the necessary measures to deal with such problems, confronting as it did the exigencies of wartime and populations adept at ignoring unwanted regulation … [and] these powerful social processes [resisting to unchanged political norms], under other circumstances, offered political opportunities' (Fforde 2011:166, 179)

This perspective can also be translated to the party–state recognition of the loss of the coercive power (the ability to apply command-and-control strategies), especially in the new Vietnamese context. The state has gradually realised the challenge of filling the gap between policy and practice which arises from diverse social processes. Additionally, a market orientation has played a large role in the transition of Vietnam since the 1990s. It has been asserted that economic factors have become more important than policy (Fforde 2010:141). Some scholars believe that evidence shows the influence of neoliberal agendas and donor pressure on policy and policy formation in contemporary Vietnam (Greenfield 1993, Masina 2006, Nguyen-Vo 2008, Beresford 2008, cited in Fforde 2011:167).

In sum, the market-oriented economy does matter, and current actions and ways of governing show continuity in some aspects of the historical relation between statecraft and social arrangements. Over the history of trial and error in Vietnam, the central state never achieved complete control over the diverse local level. The present-day state recognises and partly accepts adaptation of policies in responding to local resistance as one governance strategy. This view is proven throughout this book and supports Fforde's argument for the need to be careful in terming Vietnam as semi-authoritarian regime. The more complicated reality is characterised

> with limited, not responsible, political pluralism, without elaborate and guiding ideology, but with distinctive mentalities, without extensive or intensive political mobilization, except at some points in their development, and in which a leader or occasionally a small group exercises power, within formally ill-defined limits but quite predictable ones. (Linz 2000: 159, cited in Wischermann 2013:7).

Despite the achievement of a certain value from the grassroots democracy legislation, giving farmers the right and power to participate in decision making does not mean leaving them alone to decide whatever they wish. Thus, hamlet and commune cadres have another task: *dân vận* (people mobilisation).

> Dân vận is mobilising all groups of people, without missing anyone, together to join in a whole group of people in order to do all the tasks and mandates that the state and the mass have given. (Hồ Chí *Minh: Toàn tập, Vol. 5*, National Politics Publisher, Ha Noi)

In this context, one might define democracy as freedom under mobilisation. When used strategically according to mandate, the language promoting democracy during pumping meetings has no implications beyond the symbolic promotion of the state's image of granting people the right to decide their own business and no longer practicing the command-and-control method which failed during the collectivisation period. Although 'democracy' in this context does not imply any form of social fairness, it encourages state co-operation within local arrangements in order to effectively handle local matters and achieve farmers' and the state's target for production. In this book, it is proven that state intervention in field drainage contributes to overcoming barriers to co-operation in communal works. At this stage, the state's dilemma in guiding the pumping meeting negotiation clearly demonstrates how hamlet cadres simultaneously pursue 2 contradictory tasks: promoting the state's retreat from negotiation and maximising policy implementation through mobilisation. This dilemma is complicated by cadres' third duty of contributing to the social arrangements of which they are part as farmers.

Negotiating the pumping date and cost-sharing

At this stage, negotiation produces the pumping date and cost-sharing mechanism which, although not final, serve as the backbone for further negotiations. With farmers familiarised with the state's presence in structure and regulations and the success of the state's infrastructure power, the meeting provides the starting point for the season and field drainage activity.

Before analysing negotiation over the pumping date and cost sharing, it is necessary to note that decisions about the pumping and sowing dates are based on farmers' consideration of natural, physical and social factors. In the farmer's perspective, which is well understood by hamlet and commune cadres, the first and most important factor is the water level. Along with the duration of water retention, the water level differs by area and year, depending on the flood regime, rain situation and local infrastructure of dykes and other hydraulic works. The water level also fluctuates daily depending on the river tidal system, which guides the 2 peak and 2 low water levels over a lunar month.

In planning the sowing and field drainage for the winter–spring season, farmers always first check the water level in the canals before giving attention to other aspects, such as the state's sowing schedule. The favoured dates for pumping are low water periods – around the 6th–12th or the 20th–26th day of each lunar month. In these periods, the water in the canals is low enough to pump water out of the fields, and less water is pumped, so less money is spent. This practice is the result of interaction between human settlement and nature. Depending on the historical period, human settlements control, adapt or do both in practicing farming production.

The process of implementing irrigated, intensive agriculture with 3 rice seasons per year and making considerable achievements in infrastructure (e.g. dyke) and technologies (e.g. diesel pumps) have given farmers and state agencies the confidence to fight against nature, specifically combatting the water cycle by making better use of natural rules: using pumps during low water periods for early sowing. Thus, physical conditions are also decisive in decision-making for field drainage. Based on the condition of dykes, farmers decide to start pumping and then sow immediately or week(s) after the pumping meeting. Farmers also take into account the state sowing schedule. As a result of the state mobilisation efforts, farmers are aware of the benefits of collective sowing and the schedule's potential to avoid the hopper epidemic. Consequently, the state-promoted dates for sowing are seriously considered and openly accepted by farmers during the decision-making process. Nevertheless, the water level remains the most dominant factor in decision making. Thus, regardless of the state's people mobilisation efforts and farmers' acceptance of scientific knowledge, fluctuations in water levels due to rain in October and November (lunar calendar) can prevent farmers from practicing collective pumping and early sowing.

In addition to social requirements, technical or individual dilemmas faced by farmers affect individual and group decisions. For instance, the pumping dates must avoid celebrations or anniversaries in the hamlet (meeting 29.11.2011), giving priority to farmers who have only a single pump but several fields in separated zones (mentioned in all 13 meetings in XT and TXB communes) and intra-family conflicts in assigning labour to communal work (recorded in ĐT and TXB communes). Thus, from the planning phase at the pumping meeting, local cadres consider various aspects and overtly accept potential deviation from the state plan and targets. Since the 1990s, local stakeholders, including state cadres, have recognised local diversity and included it in the early negotiation and adaptation process at the interface of field drainage. How negotiation occurs in this early phase is analysed hereafter.

Bargaining of individual behaviour in negotiating the pumping date

Coming to the meeting with a rough decision in mind after considering the water level and dyke conditions of one's own field, farmers actively participate in the negotiation of the pumping date, regardless of their land-holding, gender and position (concluded from 17 pumping meetings in 10 hamlets in 3 communes). Observations of the negotiation in the meetings suggest several common features.

The water level and dyke conditions often pre-empt negotiations about pumping timing. No single case observed showed that an agreement could be reached when water inside the compartment was more than 80–100 cm deep at the time of the meeting. The desired water level at pumping time — at low tide — should not be higher than 40–50 cm (less than one adult arm). Without these conditions, farmers either pointed their disagreement in the meeting or were deliberately absent from the meeting to delay the field drainage process. Thus, stakeholders make use of their power to negotiate and mobilise others in order to reach a reasonable decision. This action demonstrates that the holding power of each farming household in communal negotiations, such as those over field drainage in zones, relies heavily on individual commitment to contribute. Decisions about collective works are made by individuals to benefit their private land for their private income. The individualised nature of farming gives farmers the power to negotiate and care for their own benefits, and the determinant power of natural and physical conditions trumps imperatives from the state sowing schedule or other bureaucratic mandates.

Negotiation occurs among big and small farmers, farmers and state cadres, most of whom are farmers, whether inside or outside the hamlet (Picture 5-8). Although significant, differences in landholdings, ranging from several *cong* to 100 *cong*, do not affect a farmer's say in the meeting. The farmer who has no position in the hamlet and the hamlet cadres who voice their opinions during common informal discussions are treated equally. In terms of gender, although the majority of meetings did not have more than 30% of female attendants, those attending the meeting directly performed farming and were equally respected by the other farmers. Once, a female farmer convinced the others to postpone the pumping date which most participants had first agreed upon (meeting 11.11.2011).

Picture 5-8: Discussion in a pumping meeting (Huynh Linh 2011).

However, the freedom to speak in the meeting does not necessarily mean that decisions and negotiations are based strictly on a system of equality. Many other factors influence individual and communal decision-making, including strategic field location, social prestige, the enthusiasm of farmers (state cadres or not) and state encouragement to make the event a sign of the hamlet's image as a model subjected to state reward and support.

First is the power gained through strategic physical conditions. Farmers whose fields lie at the zone's border must join the pumping as they are responsible for running the pumps longer than others in order to avoid water leaching during and after collective pumping. Their critical role gives these farmers more power in choosing and postponing the agreed-upon pumping date. These farmers are also more sensitive to water and rain levels and the dyke system for pumping as they tend to incur more fuel costs. Second, farmers whose have gained prestige through producing good farming result are more likely to be listened to during negotiation. Extra respect is given to older, more experienced farmers, though this respect does not translate directly into negotiating power. Instead, farmers with prestige from better yields and farming practices tend to enjoy more say and trust in the meetings; this respect is given, for instance, to individuals who are better at eliminating yellow apple snails and cultivating high-yield varieties. Based on this power sources, big farmers tend to have more say in the meeting. Some newly assigned hamlet cadres also possess this prestigious power relating to farming practice. In addition to socially granted power, farmers in this era of individualised farming and market orientation reserve the right and power to negotiate matters, including agreement and disagreement with other households. In addition, although the bureaucratic structure which does not apply coercive activities does not give hamlet or commune cadres power in pumping negotiation, their facilitation and mobilisation of these activities can encourage an early start in some zones.

Taking into account the state encouragement for early, on-time sowing, 2 hamlets leaders and 1 commune's party secretary in XT commune decided to make their zone sow according to the state schedule. The zone became the first hamlet to start pumping, in the second low water of October (lunar calendar, the first period stated in the state schedule). The 2 hamlet cadres made this decision with support from the commune cadre. Although there was no clear threat from fluctuating water levels, the activity was still performed early for 2 reasons. First, those cadres and their fellow farmers were confident in the zone's available dyke system. Second, intensive mobilisation actions were conducted to make this effort a pilot case or, in other words, to make the cadres successful at fulfilling their duty to be a mirror or model to the people — one of the principle purposes of cadres in Vietnam. Other farmers were convinced (mobilised) by the cadres' enthusiasm and the assurance of state support in case of failure. However, following this schedule turned out to be a mistake as the high water came immediately after sowing and washed away all the seeds sown. This event became an embarrassment for the local state, and farmers in the hamlet later used it as a reason to more carefully check the water level and to delay attempts to sow early when the water is still high. This tendency shows that the hesitation of those state cadres to take similar action in the next year and the uncertainty in farming amid unpredictable natural conditions have made local cadres shy about forcing a fixed sowing schedule, thus adjusting and adapting the state mandate to the local context. This incident also illustrates the trial-and-error nature of the process of negotiating institutions.

Finally, the social ties between neighbours also play a role in negotiation. While expressing one's view of whether the zone's sowing will be too early or too late is important, some farmers decide to compromise with others in order to reach a communal decision. Farmers tend to hesitate to object agreements during meetings, perhaps to avoid losing face, reluctance to speak in public (many women tend to whisper in small groups, rather than address the whole audience) or uncertainty about the reasons for the objections. As well, farmers favour early sowing to gain better prices (rice harvested early or later tends to earn better prices than rice harvested in the middle of the season), creating an imperative to reach an agreement during the pumping meeting. Although the decision made in the meeting is not final and is constantly discussed and adjusted in public community spaces, stating one's view at this beginning stage is important as the meeting is part of the whole negotiation process. Thus, the selection of action during the meeting — whether to raise an issue or not, to discuss the issue immediately or later, to object to the agreement or be too embarrassed to do so during the meeting — requires complicated consideration of bureaucratic, economic and social factors and is in constant negotiation with individuals' socio-economic status or holding power. In the researched communes, large farmers noted for good practices tend to be active in the negotiation and to express objections, while small farmers, including many women, do not officially state their objections. The latter group waits for further informal discussions outside the meeting, supports anyone who raises the issue during the meeting or completely withdraws from the pumping event and do pumping their own way. Above all, farmers retain the veto power to withdraw themselves from collective field drainage as farming has been an individual matter since Doi Moi.

In all observed hamlets, farmers and local cadres reported difficult households which have been more or less stubborn – or free-riding problematic or individualised farmers holding themselves independent from the social ties. Intra-household labour shortages, conflict among members of a household, past conflicts with neighbours and bad sentiments towards local cadres' management and facilitation methods are amongst the reasons for self-isolation. In 1 case, the feeling of inequality resulting from state management of land use de-motivated a household from engaging in communal co-operation. In another case, the intentional free-riding of a household forced other farmers to close the dyke section of his field in order to start collective pumping.

In sum, no elite group holds absolute power in the contemporary Mekong Delta. Rather, the power to influence the negotiation process of pumping dates and cost-sharing mechanisms arises from multiple sources and varies in its effect. The evidences imply a pattern of individual behaviour or decision making towards co-operation that confirms the following statement.

> People can purposefully plan their actions in relation to the collective, but this is likely to be informed by 'emotional', 'moral' or 'social' rationalities as well as economic one. (Cleaver 2012:15, based on Scott 1987, Boelens 2008)

Positive or negative sentiments towards a person, agency, way of organisation, the ideology of living at peace with the neighbours, the availability of resources to contribute and the economics of joining communal co-operation are all taken into account and affect farmers' behaviour. The influence of each factor differs among individuals and activities and is affected by the natural, physical and political circumstances. In the case of field drainage, the economic rationality or economic interdependence among farmers facing a common threat (Howie 2011) becomes the most important driver of collective action. This trend is especially true when interdependence among farmers is low. For instance, farmers from the rice–fish culture are independent of dyke conditions for field drainage and so of collective pumping. In the contemporary Mekong Delta, the individual characteristic in farming production — the winter–spring production is returned to individual farmers after collective field drainage — gives little scope for favouring moral, social and even emotional rationalities over co-operating in collective activities. The further exploration of the field drainage process applied in this chapter provides evidences for the pathway to action and helps clarify the diversity in the extent of these factors' influence on individual behaviour.

Fuel sharing: The clash between social cohesion and individualism in farming

> With a pipe of 22 cm diameter, the cost for each hour [pumping] is VND 30,000. Each 2cm bigger in pipe diameter will add VND 2000 per hour. [The total cost will be calculated for the whole zone]. The household having more land will compensate the cost to the household with less land. The group leader calculates the money. (hamlet cadre)
>
> The village's love and the neighbourhood's relationship (Tình làng nghĩa xóm) — We are all neighbours, what for to calculate? … We will calculate together in small groups, and by ourselves share with each other [either by fuel, money or in kind such as party]. (experienced farmers, former Farmer's Union cadre at hamlet meeting)
>
> (Pumping meetings, 29.11.2011)

These 2 contrasting expressions were recorded in 2 pumping meetings in the same hamlet. They hint at the current diversity in local arrangements, individual perspectives and changing tendency in fuel sharing for pumping.

Once the pumping date is agreed upon, hamlet cadres discuss the second topic of pumping meeting: fuel sharing. As mentioned, the pumping stage of the field drainage activity has 2 steps: first is collective pumping until the water level falls just below the ridges separating fields (until the local ridges are seen); second is individual pumping when households take care of their own field. Farmers call the second step the period of '*catch your own chicken*' ('*gà ai nấy bắt*' – interview 26.10.2011). Not all households have a pump, and farmers have differently sized land-holdings, so the fuel used in collective pumping has to be calculated and shared. Fuel sharing is categorised as an action of communal sharing to achieve the fairness which requires as much monitoring as those involved feel necessary to gauge fuel usage. In contrast to deciding on sowing, hand labour or in-kind contributions, fuel affects the negotiations as a strict monetary. However, as demonstrated in this book, various methods to share the fuel are applied, and many of these are inaccurate and tend to be similar between places. Whether or the fuel calculation is accurate, the action of positioning the issue in the meeting fulfils the social requirements for collective work, which differs across pumping groups. While all stakeholders agree that the farmers attending the meeting decide on cost sharing, state cadres are facilitators and arbitrators of conflicts (e.g. not bringing pumps, not pouring the fuel during pumping).

The most common method of fuel sharing is to estimate how many litres of fuel (commonly 2–3 litres in the researched communes) are needed for 1 *cong* (1,300 sq m). According to this method, households which do not contribute pumps and households with more *congs* have to give back fuel to farmers who contribute pumps and have less land. A more complicated way also takes into account the diameter of the pipe attached to the pump. In Xuan Thang commune, it was suggested and discussed that the owners of the smaller-pipe pumps compensate the owners of pumps with bigger pipes.

Fuel calculations are local arrangements not necessarily based on accurate economic calculations or equality. The exact amount of fuel used in pumping varies annually depending on the water level, the condition of the covered dyke and any unexpected prolongation of the pumping period due to a sudden rain or dyke damage. However, the amount of fuel must set in advance as the first step of proving co-operation and starting the fuel-sharing process. In the meeting, farmers agree on a method of back payment to ensure that all contribute equally to collective pumping. However, the practice of equality is adjusted to the specific social arrangements of certain groups of farmers and challenged by free-riding actions and the uncertainty in fuel usage, which might change significantly due to fluctuations in water and rain levels and dyke conditions.

At meetings which calculate fuel costs based on *congs,* agreement was often reached easily for 2 reasons. First, all have agreed on the shared method over the years, despite the uncertainty of its ability to promote equality. This social acceptance of certain local arrangements due to their popularity or the inertia of institutions, instead of their value for equality or social fairness, is shown in the case of calculating contributions for dredging field canals. Second are the culture of sharing and the relationships among farmers in the same community. In that sense, strict accuracy in fuel or money calculation is not common or necessary amongst neighbours. In the market-influenced community, *village love and neighbourly relations* are still valued, but the extent of their influence depends on cases and localities and is in constant negotiation with the new, individualised lifestyle, which is busier and has more difficult in sustaining life's requirements.

However, this modern lifestyle also puts pressure on and forces the modification of the social structure of continuous, in-kind sharing. When pipe diameter is also taken into account for better accuracy, the group leaders or hamlet cadres acting as facilitators note the available pumps and pipes used by each household and calculate the money that households will give or receive based on their land holding. This pattern clearly shows the local incentive to move towards a more accurate or fairer way of sharing amongst farmers. The selection of a more accurate sharing method arises from the persistence of unfairness in old sharing methods (described in the section on the implementation of fuel sharing) and the increasing difficulties in finance and social mobility that challenge the social system of continuous, in-kind sharing within communities.

Despite the desire for fair fuel calculations before and after pumping, direct cost sharing was not practised in some cases in the researched area. More than 1 particular way of sharing in the community was found. Some farmers even favour not discussing the matter as '*it is weird to pay each other* [to pay neighbours]' (interview 31.11.2011). One example is as follows:

> In THB hamlet, Xuan Thang commune, it was told in a pumping meeting that a farmer with 10 ha of land did not need to pay back the fuel to his 6-hectare neighbour because they are good neighbours. They have shared a lot of other things throughout the farming process, such as surplus materials, fertilisers or pesticides; or the big land holding one just simply needs to set a party after all. (interview 21.12.2011, 22.12.2011)

In addition, joining and contributing towards eating and drinking in the field during the pumping period eliminates some differences in fuel spending. Farmers measure fairness in co-operation not monetarily but as the practice of maintaining *the village love, neighbourly relations* and established institutions. For those farmers, monetising relations is socially awkward. This example gives proof of the coexistence of remaining social ties — neighbourly relations — in Southern communities and the new, more accurate methods for assessing monetary contribution in changing context, though their implementation is uncertain.

Nevertheless, the culture of mutual help in the community does not apply to all cases but depends heavily on the relations among farmers in a zone and differences in land holding. Mutual help is more popular in zones where farmers are relatives or have been good neighbours for a long time as kinship and social cohesion play roles in communal work. This tendency leads to scholars' conclusions about social cohesion and kinship value in the Mekong Delta. Although the Delta's open society has looser kinship network than Northern villages (Hy V. Luong 2003:23, Hy V. Luong 2005:141), the long period of domination by Chinese culture, with its Taoism and Confucianism ideologies, has influenced Southern Vietnamese villages, including their kinship network (Hickey 1968:5), so that '*Vietnam is a kinship society, not a civil society*' (Rambo 2005:7). Today, the smooth operation of fuel sharing relies on human tolerance of social unfairness as fuel sharing has not been calculated accurately yet, and in-kind compensation from other parties and mutual help in everyday life are not available or valued similarly by all (these issues are further explored throughout this chapter).

In sum, there is no clear marker of what arrangement a zone will favour. Instead, the 2 options are in constant negotiation for every individual. This situation gives rise to diversity in decision making and selection among individuals and stages of the pumping process. The mixture of methods which follows no particular pattern illustrates the pathway or process to form, choose or legitimise an institution at this specific period.

Meeting results: Signed but not final agreements

Picture 5-9: Signing of agreement at the pumping meeting, the minutes and signed agreement (Huynh Linh 2011)

With an agreed-upon pumping date and method of fuel sharing, all meeting participants sign an agreement as a proof of their consensus. However, the decisions made in the meeting are often not the final, especially in years when the water level changes unpredictably over time, as in 2011. Thus, the signed agreement or contract is a symbol of good faith but is not actually so because of the need to adapt to changing circumstances. Adaptation is made in after-meeting interactions and negotiation in community spaces, such as markets, households, fields and roads; as such, it is called everyday negotiation. The recognition of a 2-step process implies the bureaucratically oriented state's acceptance of flexible implementation. The following exploration of the pumping process provides more explanation and analysis of these issues.

The following narratives describe what might happen after the agreement is signed. The 2 cases of the zone BC-UC in Truong Xuan B commune and of zone 1 of Xuan Thang commune demonstrate the process of negotiating the pumping date during and after the meeting. In many cases, postponement by some farmers causes a delay in achieving full participation and sometimes changes in the zone boundary or the elimination of the fields of the farmer who caused the delay.

The first challenges: waiting and cutting zones

When practice is started, ideal plans are subjected to changes due to uncertainties, including climate fluctuations, unstable physical conditions and continuously changing institutions for local co-operation between farmers.

In zone BC-UC, Truong Xuan B commune, the third pumping meeting took place on 17 October (lunar calendar). In the meeting, participants agreed to pump-out on 22 October (lunar calendar). Repairs to a wide opened dyke section delayed pumping by 5 days, and a lack of labour caused further delays. However, at that time, the pumping involved only the first half of the zone. The second half only proceeded almost a month later, on 21 November (lunar calendar). The deviation from the agreed-upon implementation did not create much pressure because the middle strong dyke enabled zone splitting. After negotiations with farmers on the other side of the dyke failed, the farmers decided to narrow the pumping area. Thus, depending on natural and physical conditions, tensions can be dissolved, and adaptations quickly made. Quick responses reduce the tension in non-collaborative situation and create a higher chance for continued selection of individualism (including free-riding actions) over communal co-operation as the risks of breaking neighbourly relations are actually low.

There was not so fortunate a resolution in another case in the Xuan Thang commune, when a group of households had to delay their pumping schedule while waiting on farmers in the same compartment. When the water became lower than the alternative dyke, these households decided to cut out the hesitant households and start pumping in a smaller zone. In this situation, the tension was high between farmers, and some developed resentment at the difficulties in co-operation. This situation could create a foundation for institutional change or new bricolage.

The dyke system is a symbol, if not the zone boundary, of the mutual interdependence of stakeholders within a connected zone. It acts as either a main driver or a barrier to co-operation in communal work. If a zone and interdependence correspond, they can act as an effective bureaucratic and economic unit, which favours the imperative of state management at the local level. Overall, physical conditions, particularly the dyke system, play an important role in deciding the results of co-operation in field drainage activity. In the second case, the dyke condition which prompted high tensions in ae situation of in-co-operation added to the incentive for using contracted pumping as an alternative because paying the exactly same amount of money for a certain amount of contracted land creates a sense of monetary equality amongst farmers. Therefore, the forming of bricolage shows diverse characteristics among areas as institutions are locality dependent.

Preparing for pumping: Gathering the pumps and repairing the dykes

Although dyke repair is vital for the start of field drainage process, it commonly faces obstacles to arranging communal contributions which are temporarily resolved by individual goodwill or tolerance. The following 2 examples illustrate the constraints that appear during the preparation stage. The failure of negotiation to mobilise a workforce for collective work is compensated by the tolerance of individuals who perform the work all for the communal benefit. Differences in the need for pumping, vulnerability, individual interests and consideration of financial, social and bureaucratic situation make the outcomes of negotiations over dyke repairs not guaranteed. These differences continually present the risk of certain individuals taking on the burden of collective activity. Those more likely to take on the burden include hamlet cadres responsible for the progress of state and local planning, enthusiastic farmers with high social spirit and the most vulnerable farmers who cannot start early farming in their fields without collective pumping. Dyke repairs, then, lie at the intersection of economic vulnerability, interdependence, state encouragement and voluntary spirit.

Dong Thuan Commune

In group 3, hamlet ĐH in Đong Thuan commune, which has 208.5 *cong* and 20 households, farmers held a pumping meeting on 24 November and planned to start pumping on 30 November (6 November lunar calendar). However, the high water level forced the pumping date to be changed to 4 December and then 5 December to accommodate dyke repairs. Observations and interviews illustrate the possible failure in mobilising the labour force and contributions to collective activity.

Upon the agreement to change pumping to 4 December, all households set their diesel pumps on the bank of the compartment. However, pumping could not start because of the need for repairs to one large dyke section at a critical position. Other farmers were forced to wait while zone leaders attempted to persuade the family to repair the dyke and join the pumping. Unfortunately, an agreement could not be reached. All the pumps were in the field, and a long waiting meant longer field work for all farmers to avoid losing pumps, so the group leaders and some farmers repaired the dyke section. The repair work was completed in the morning of 5 December, and the pumping started shortly before noon. The difficult household that did not join dyke repairs and pumping risked losing social status and cohesion with neighbours. The 7-year repetition of this issue had already created social distance between the household and others, and the resulting negative sentiments and bias on both sides made the negotiation even more difficult. Thus, past experience and the lack of social authority — no one has more power in negotiating private farming practices, not even the zone leader — created constraints on local arrangements of co-operation. The resulting solution relied on the tolerance or goodwill of some individuals to take over the work of others.

The existence of negative sentiments alongside the goodwill is clearly illustrated in the following events of the in-co-operation of the difficult household. In the early morning of 5 December, one farmer, annoyed by the actions of that household, took his pump back to his house. He was convinced of the benefit of starting the pumping as soon as possible, though, so in the end, he brought his pump back to join the group. These actions are an extreme case of popular response from farmers when inequality is felt.

At the Truong Xuan B Commune

Picture 5-10: Dyke repairs before pumping (Huynh Linh 2011).

The wide-open dyke section must be closed by all farmers in the same zone. Tree trunks donated by individuals are used to make the frame which will later be covered by a net or plastic. The final step is to fill the structure with soil, which can take up to 2–3 days. In the picture, trees are planted to frame the dyke, which will be made from soil covered with plastic.

An extremely common problem in the preparation stage is that few farmers handle the main dyke section which should be performed collectively by all farmers in the pumping zone. In the Zone BC-UC, Truong Xuan B commune, the pumping date was delayed 3 days while waiting for dyke repairs. An approximately 20-m dyke section was completely underwater and needed to be repaired or rebuilt for pumping to start. More than 10 farmers (of 33 households in the zone) agreed to meet at the section 1 day before the agreed-upon pumping date to collect trees and (re-)build the dyke. The researcher and 2 hamlet cadres arrived first. After waiting and calling other farmers, 4 arrived approximately 1.5 hours later. The work finally got started (Picture 5-10) and reportedly took 3 days to complete; on the second and third days, only 2 hamlet cadres and 1 farmer continued the work.

The main reasons farmers gave for their reluctant behaviour were hesitation to face high water levels and potential alternatives for gravity drainage if the water receded quickly enough at the next low tide. The refusal to contribution or help creates disappointment and the feeling of inequality among other farmers. Consequently, they leave the work to hamlet cadres, (former) mass organisation cadres and others with high social spirit. Current and former hamlet cadres generally have a high social spirit, which is a reason for their selection as cadres and fulfilment of a cadre's social duty to be responsible and a model for the community. Thus, within the hierarchical state system, bureaucratic pressure plays a role in local organisation which supposedly is the duties of farmers.

These 2 examples show that co-operation between farmers plays a decisive role in the preparation stage of field drainage. Social co-operation among neighbours has become difficult due to costly, busy modern lifestyles, which supports calculating the monetary cost–benefit of taking care of one's own household against communal benefits. The survival of communal co-operation in the researched area today depends on the state structure of organisation, with cadres encouraging, mobilising and keeping on track local co-operation; individuals who have the goodwill or high social spirit to do good for the group, even at a cost to themselves; and others who tolerate non-collaborative behaviour in order to get done the work for the season. Although those with goodwill and the state structure were able to complete the collective work in this example, various responses from fellow farmers, from complaining to taking action, cause uncertainty and prompts the search for more equal alternatives, arrangements and institutions. This tendency is further seen in the process of field drainage.

Setting the pumps and pumping

The stage of field drainage illuminates 3 important aspects of rural institutions: the careful, individual calculation of costs before starting work, the belief in the strength of collective force and farmers' experience in risk-taking decision, and the existing space of social cohesion and sharing despite the individual orientation and institutional bricolage.

The risk of increased from interruptions to pumping forces farmers to continue pumping until the task is completed. As a group leader tells:

> About 4–5 years ago, it was raining just after the pumps were brought in the fields. The water level was even over some covered dyke sections. In that situation, we decided to just strengthen the dyke and pump as scheduled. Nobody wanted to bring the pumps back home. (interview 19.12.2011)

Consequently, individuals carefully consider whether to start pumping. They do not bring pumps to the fields until they are have sufficient confidence in natural and physical conditions (dyke conditions, rain, water levels) or encouragement from cadres, which include promises of support.

After the dyke sections are repaired, farmers are required to bring their pumps and set them on the bank of the compartment 1 day or at least hours before the pumping start as a sign of equal contributions. In addition, the tradition of raising a flag to let all farmers know that they should start their engines at once (practised in some communes) sets the clearest expectation of co-operation and is a symbol of communal action. On-time pumping depends on willingness to start pumping on the agreed-upon date and household resources, including pumps and labour. The first challenge persists as decision making is based on individual consideration of various factors (both external and internal ones). However, the limitation of pump availability has been reduced by changes in rural areas that allow each household to have at least 1 pump. Adding to the motivation, farmers with a sense of competition follow the group's initiative in pumping early and, thus, sowing early. In some cases (in commune Truong Xuan B), the start of the pumping engine with the well-known sound of '*đùng đùng*' is expected to encourage hesitant farmers to bring their pumps and join the effort (after meeting 12.11.2011).

Throughout the field drainage process, there is an element of risk. Amid uncertainty, farmers are willing to take risks with relative confidence in the benefits of communal co-operation and farming experience. Natural conditions, such as water levels and rain, are unpredictable, so waiting for certain clear weather would mean that pumping could never begin. The willingness to take risk indicates the belief that neighbours can overcome some natural limitations by co-operating with each other. For instance, a little rain will result only in little more time for pumping if the group does extra pumping. Likewise, some farmers believe in the strength of collective action and farming experience in deciding whether to deviate from the state sowing schedule. A former hamlet cadre states:

> The scientists recommend that we shouldn't sow on the 10th [(lunar calendar), the middle period between 2 sowing period of the 2011's schedule]. Well, we farmers have lived and are familiar with the hoppers. We don't have any fear, right! (pumping meeting, 29.11.2011)

The farmers are familiar with a solution for the hoppers: Cover them with water. Hoppers are sensitive to water, so submerging the seedlings in water can prevent further damage from hoppers. Farmers also have a list of pesticides to use on hoppers.

Pumping continues day and night and is an occasion for neighbours to gather to discuss farming practices and strengthen their relation. The duration of pumping varies from 1 day to 1 week, depending on the water level, weather (any rain while pumping) and dyke condition (good, leaky, broken). The diesel pumps can run alone for days as long as fuel is supplied, but farmers have to stay in the field to supply fuel, watch their pumps and provide help in case of dyke damage. Often, rice fields are away from houses, so farmers construct small houses or tents at the field site to live in while doing farming activities, including pumping (Picture 5-11). Those field houses act as public spaces for neighbour gatherings for food and drink immediately before, during and after the season. In the researched area, the gathering continues throughout the season as a form of social interaction and knowledge-sharing. Here, in the field, the individual style of living and farming in the contemporary market-influenced era is less dominant, leaving space for institutions that promote mutual help and sharing among farming neighbours. Contemporary rural institutions are a mixture of arrangements which might be dominant in one interaction but not another. This mixture of old and new features of social arrangements indeed illustrates institutional bricolage as conceptualised by Cleaver and other authors (see Chapter 2).

Picture 5-11: Field house in the rice field (Huynh Linh 2011).

Fuel sharing: 'Counting each coin' vs. 'there are many other ways to pay'

The researched area presents a variety of options for fuel calculations and sharing, depending on the social pressure of imposing equality and the traditions of mutual help in the community. Therefore, the implementation of the agreed-upon method to share fuel (if an agreement could be reached in the meeting) varies among places. Difficulties can arise in imposing equality and the selections of methods of sharing other than accurate monetary calculation.

First, deviation can occur because delays in compensation create difficulties in implementing monetary equality in fuel sharing. This deviation creates negative sentiments among zone leaders with high social spirit, which is the foundation for the search for and support of alternatives in field drainage. As mentioned, fuel sharing in field drainage follows 2 different patterns. One, the calculated fuel costs (based on certain estimations, including the number of litres per *cong*) are shared among farmers participating in the pumping event. In some zones, the exact amount was calculated '*to each coin*' ('*từng đồng từng cắc*') (interview 29.01.2012). However, the zone leaders responsible for this sharing generally had difficulty monitoring the sharing process. Some farmers were late in bringing fuel to compensate the other farmers who had more *congs*. Consequently, leaders often first had to handle the fuel sharing at their own expense and then try to recoup them (interview 29.01.2011). That difficulty has created uneasiness among group leaders, especially when combined with the large workload they undertake without any compensation. Second, farmers sometimes choose to compensate others for fuel through other forms of social sharing. In the researched area, in-kind sharing is also practiced by farmers who believe that it is '*weird to pay each other*'. The older, more inaccurate calculations of pumping fuel will likely remain as long as the social spirit and tolerance of some individuals allows it. The *điệu nghệ* lifestyle (to live with heart and help neighbours, see Son Nam 2003) built in the Mekong Delta continues to leave its imprint upon modern society.

The selection of the counting-each-coin strategy indicates a desire for fairness. In this situation, fairness is expressed in monetary terms as farmers have less time and opportunities to engage in different kinds of sharing which can compensate for fuel differences. However, accurate sharing faces the challenges of free-riding action and individualism, which might be caused by household differences in financial capacity, costly lifestyles and individual behaviour. In addition, arrangements vary by place and group of farmers as diversity causes different perspectives of monetary calculation among neighbours. In a complex institutional process, certain arrangements are selected from the diverse materials available to form a bricolage that suits specific conditions. This legitimation varies among places, making it hard to predict and form unified rules. Therefore, crafting regulations (Ostrom 1990) for institutions to achieve equality is not possible.

Monitoring and crisis management

The pumping process has various risks, so monitoring is essential to avoid interruptions due to broken dykes and sudden rain and to ensure fairness among participants. Monitoring is followed by crisis management, in which zone leaders and assistants often encounter the regular difficulties of a lack of authority or social power to mobilise contribution. The continuous pumping event, though, creates a critical, shared situation of economic and social interdependence which activates farmers' social spirit and helps avoid individual losses.

Zone leader(s) carry out the monitoring task by taking 2 or 3 walks during the day and 1 or 2 walks a night around the covered dyke (where farmers and their pumps are based) to detect any dyke damage and free-riding actions. Some free-riding actions recorded in the researched area are deliberately turning off a pump or adding little fuel so that the pump stops in the middle of the night. When such actions are detected, zone leaders take immediate action, talking to and persuading farmers to start pumping again and adjusting fuel calculations.

The effectiveness of free-riding actions depends on the social prestige of the person committing them as the zone leader position conveys no authority over fellow farmers. The lack of power possessed by the holders of the state-designed position of zone leader reflects the historical and current context of the researched area. Social arrangements in the private farming system are regulated by local institutions shaped by the local structure of prestige and economic-based cost–benefit calculations.

The effectiveness of monitoring free-riding is uncertain as Southern Vietnamese communities have no absolute social leader, and uncertain social reactions likely can be avoided with tolerance. Despite state facilitation, a pumping group or zone is formed by farmers connected in an existing structure of local institutions with a mixture of inter- and independence in farming. Southern farming society does not provide authority to any individual, even the group leader. Instead, pumping groups are based on interdependence (especially economic), equal rights (in participation, negotiation, decision making), unequal duties and a lack of central power. Therefore, the only constraints on free-riding behaviour are economic interdependence, social pressure and strong reactions by other farmers to the inequality. Indeed, social pressure from other pumping participants can have an influence. For instance, some recalled a strong response to free-riding action as a warning: '*Whichever pump stops working* [on purpose], *we will kick it to the water* [...]' (pumping meeting 11.11.2011). Although the threat to kick pumps in the water received a look of hesitation and little smiles from the meeting participants as if to say that was not necessary among neighbours, it does indicate the possibility of strong reactions and social sanctions in the field, especially amidst the critical time of pumping. However, the uncertainties of fairness and the increasingly costly lifestyles have challenged individuals to maintain their high social spirit and tolerance. Consequently, alternatives for fairer arrangements, such as service–provider and customer–contracted pumping) are fast entering favour. This trend also indicates an increasing loss of trust in collective action within market-oriented farming community.

While walking along the dyke to check the pumps, the leader also looks for any dyke damage. If there is,

> The small one, I just fix it. If it is big, I phone the others who are staying in the field with their pumps to come to help. ... It is easy to mobilise the labour [at that time] because if we don't fix it, we have to pour more fuel. (group leader, interview 29.12.2011)

As stated, the response to dyke damage during pumping is fast and involves all farmers involved in the pumping. The availability of the farmers staying in the pumping zone makes mobilisation easier during crises, such as a broken dyke. Also, at the pumping stage, all farmers face a common threat, which motivates collective action. If they do not join repairing the broken dyke, fuel costs will be higher for all. When obvious and perceived by all, economic interdependence and mutual vulnerability create great incentive for co-operation and collective action. In these moments, the common benefit and the individual benefit become aligned.

In addition, such situations also activate and support farmers' high social spirit. Once, while waiting for others to come to fix a broken section of dyke, a farmer in Truong Xuan B commune farmer pushed his boat over the gap to stop the violent flow. The action not only demonstrates behaviour to address a common threat but also the voluntary action of an individual with high social spirit. Although any farmer can commit acts exhibiting high social spirit, current and former state cadres, assistants and associates are most likely to perform them. Thus, the pre-Doi Moi social engineering (especially for cadres) and the persistent ideology of cadres as models still help overcome communal hurdles, despite the rise of capitalism.

Day-to-day decision-making and adaptation

Increasingly overtaking the clear-cut, technocratic and bureaucratic orientation in the management of the Mekong Delta waterscape are more decentralised, informal and day-to-day forms of decision making that tend to be more adaptive. In the power vacuum between the very local cadres and residents (such as in pumping zones) opened since Doi Moi, this type of decision making is growing in importance and has been unofficially recognised by the upper-level state. Here discussed are several aspects of the everyday decision-making, negotiation and adaptation in field drainage. These demonstrate the continuous negotiation as an activity progresses from the public sphere to inside the household and from official meetings to informal social spaces.

Intra-household decision-making

Like any farming decision, those about field drainage are intimately affected by intra-household negotiation and decision-making. Often, only 1 household member comes to the pumping meeting(s), most often the man considered the head of the household. However, other family members also have a say in decision-making. All interviewees agreed that, when both a husband and wife farm together, the information received in the meeting needs to be discussed back with the one who is not there, and the decision-making is a joint process. For example, '*when the husband attends the training on a new sowing technique and agrees on practising it but later discusses it later with his wife who is not in favour of the suggested changes, it will not be applied*' (interview 10.11.2011). This pattern calls into question the utility of meetings attended by only 1 family member and the legitimacy and credibility of decisions made and consensus reached at such meetings. Also, the concept of a household head in the farming households of the Mekong Delta is questioned. The following case study features a family system co-existing with a rigid, structured meeting and social dynamics.

An in-depth conversation with a female farmer which is also Women's Union cadre and observation in the pumping meetings shed light on the causes of this structured labour division and decision-making within families. After the pumping meeting, we met again at her house where she and other farmers in the same zone were waiting for the water to fall. She had moved to the commune, her husband's hometown, and stayed and farmed there for 22 years. The family has 5 *cong*, and both she and her husband participate in farming. She describes how work is divided between the two:

> Both wife and husband go farming like this: I am doing weeding while he is spraying fertiliser and pesticide. … About seeds [the rice variety] and the sowing time, I let him decide. I let him have a bit more power. … In general, each brings 'a hand and a leg' [to contribute to the farming], understands each other, so, it is easier. In general, we divide; each one has his/her own task. To U [her husband], I decide to 'go 1 step after' [give him more power]. Anyway, he is the head of the household. We have been together for a long time and have no problems at all. One talks, the other listens, and the other way around. (interview 01.12.2011)

Regarding who attends the pumping meeting, she states:

> Most of the time, the men who are supposed to know more about dyke and pump [they are the ones who operate and build the dyke] will join the pumping meeting. We, the women, when we go, we can also give opinion, but they [the men] know better. … In general, with work that we both know, either can go to the meeting. (interview 01.12.2011)

Since the informant was busy with the organising work of a hamlet cadre, her husband went to the pumping meeting in their zone. However, they both know about farming, so she joined him after the meeting to make decisions. Although, from the outside, the arrangements in intra-household decision-making appear that only men make decisions, eligibility to participate is determined by participation in farming. Once individuals work in the field and take care of farming matters, they have gained knowledge and developed concern for important matters in deciding the harvest.

In addition, the division of men's and women's work and the tendency for women to stand 1 step behind men in public illustrate the continuity of the historical culture of family and gendered society in Vietnamese. As stated, intra-household task division designates men's work, women's work and joint tasks. While men are more likely to take responsibility for technical know-how, such as pump operation, women take care of less heavy work and household matters (i.e. taking care of the house and children). In the researched area, the family model in which the man takes care of farming while the woman takes care of household matters is dominant. A typical picture is that the husband handles farming, while the wife takes care of parents, children, grandchildren, household matters and perhaps a small business, such as cafeteria (short stay in 12.2011). In this case, the husband comes to the meeting, performs pumping tasks and decides the date of sowing and related issues. The tasks are split between husband and wife who each take on specific responsibilities.

Although the example might not hold true for all cases in the rural community, it illustrates the existing task division and negotiation within household which is often not visible in public places, such as the pumping meetings. In short, in a rural household, tasks are divided between men and women. While men perform heavier tasks, such as dyke repairs, spraying and staying overnight for pumping, women are responsible for weeding in the field, household meals, rearing animals, fishing in the flooding season and collecting yellow apple snails for sale or for household net aquaculture (Picture 5-12). The gendered division of tasks between men and women is not specific to the Mekong Delta but is also recorded in Northern Vietnam, manifesting the adage: '*The husband ploughs, the wife transplants, the buffalo harrows*' (*Chong cay, vo cay, con trau di bua*) (Bergstedt 2012:6). Bergstedt (2012) captures the gendered labour division in a more general sphere, recording the view that men do 'big work' and women do 'small work', which can easily be found in the rural Southern Delta. In the Vietnamese context, gendered task division is the result of historical influence, especially the cultural principles of Confucianism. The '*Confucian ontology is not based on equality but on an ordered hierarchy where each person respects her or his position in the social structure*' (Bergstedt 2012:145). In this system, maleness is associated with *duong* (the Chinese *yang*), heaven, movement, hot, superiority and the positive, while femaleness is associated with *am* (the Chinese *yin*), earth, passivity, cold, subordination and the negative (ibid). Consequently, men are attached to a masculinity which possesses greater analytical abilities, harder bones and stronger bodies than women and thus is responsible for heavy work and decision making in the household and community (ibid:124-125). In contrast, women are more '*light, easy, and delicate, …* [and the] *capabilities of endurance and patience … and being careful and painstaking in their work were acclaimed attributes of femininity*' (ibid:125); thus, women are responsible for delicate work that men's clumsy hands cannot handle. Popular opinion and belief long held that the division of tasks between men and women is natural according to the given capacities of each sex (ibid:178).

Picture 5-12: Top: Dyke repair as heavy work – men's work; Bottom: Transplanting as work requiring patience – women's work (Huynh Linh 2011).

However, these traditions and institutions are changing as women take on some of men's tasks. In most cases, rural women farm together with their spouses, giving them power in discussing certain farming issues, though their level of power can differ from their husband's, depending on the task division and intra-household arrangements. These cases also raised the issue of women's choice to give men credit in public place as representing the family and making decisions on issues (e.g. pumping dates) for harmony in the family. These actions illustrate the persistent belief in the naturalised system of gendered labour division. These actions suggest 2 possible interpretations. First, the action of freely offering power to males happens only in the case of wives with little role in farming. Second, more frequently in households where both males and females do farming, women's step backward to give males space in public is a soft strategic strategy to achieve household harmony while still holding power in negotiating and joint decision making on farming issues.

Local communication networks: Cafeteria and everyday encounters

In addition to individual household decision-making based on intra-family negotiation, decisions are significantly affected by the information flow and negotiation in public spaces and through everyday conversations with neighbours. The social network has been extremely powerful in encouraging either achieving social consensus or rebelling against the decisions made in the pumping meeting.

Whether scheduling decisions are made in or after the meetings, the flow of information is important. Observations, including short stays in the hamlets, help to understand the influence of after-meeting discussions and public encounters or, in other words, everyday negotiation and adaptation. Negotiation is not restricted to official meetings; rather, it is imbedded in everyday interactions. Decisions might be changed over time through the fast circulation of information and the continuation of post-meeting negotiations in public places, such as rice fields, markets and cafeterias. Below is an example of events in public negotiating places.

> Xuan Thang commune, morning 22/12 (28/11 lunar calendar): Until today, part of the hamlets had rice plants more than 3 weeks old, some other fields had 2-week-olds, and some parts had just finished pumping on that day. It was in the house of 1 of the group leader which was, at the same time, a cafeteria. Around 5:30 am, the first guest came for coffee. The number of farmers increased to around 10–15 from 6:00 am. The owner of the cafeteria, the group leader, also joined others for a morning black coffee, a common habit shared between men in the rural area. On this occasion, we could see familiar faces whom we had interviewed and many others. They talked about what happened in the hamlet,

> in the commune, about celebrations, anniversaries and, of course, the pumping, sowing and rice season. Information about when he pumped, when she sowed, when they eliminated snails, how much I paid for fuel, how the plants were growing, etc. were excitedly discussed in that everyday meeting. Discussion also focused on the households who just finished the pumping-out on the day, how they hesitated in facing the water.

The leader's wife, who ran the cafeteria, stated that the number of farmers there that day was small. Immediately before the pumping date until sowing was finished, the cafeteria was crowded every day with the farmers who came to have a cup of coffee before going to the field. The cafeteria became an information point and negotiation space during the pumping and sowing period. To many farmers, coming to this place and discussing matters was an everyday habit; as one said, they become '*long-term customers*'. This habitual, established institution which is context defined continues to have a strong influence on the negotiation of practice and the legitimation of previously discussed arrangements.

Another public place for farming negotiations is the rice field. Farmers observe their fields regularly and there meet and talk with other farmers and local cadres. In February 2012, the researcher joined the Farmer's Union cadre and a former hamlet cadre for a field observation when the rice fields were ripening. When we were at the junction of various pumping zones, a group of farmers approached us and started asking the Farmer's Union cadre about the rice season, sales opportunities and the canal dredging plan (Picture 5-13). The informal conversation went on with explanations from the cadre and suggestions and discussions from the others. Although there was no urgent decision to be made then as in the pumping period, the encounter provided a space for receiving up-to-date news and colleting opinions and feedback. Additionally, the speaker supplied information to the farmers. Information transferred through speaker varied from elections to weather forecasts to successful agriculture models. While the speaker engaged in one-way communication and supplies general information for the whole commune, it was complemented by other social networks for information distribution.

Picture 5-13: Encounter of farmers, hamlet cadres and researcher in the field (Hai Trieu, 2012).

In the rural Mekong Delta, farming households are scattered along the canals, and life follows the water flow, so one farmer often knows the issues of other families living kilometres away. The water network connects farmers, both symbolically as the flow of water — the flow of a network — and materially as the connection among farming households for transportation, collective pumping, farming information and technology exchanges. Over time, rural inhabitants have formed an effective communication network through which issues are related, discussed and negotiated in informal encounters amongst neighbours. These local institutions are maintained by and vary among localities. The neighbourhood or the institutions for social arrangements between neighbours in the rural Delta have been influenced by internal and external factors and thus subjected to changes which, in turn, are contextualised by the locality. Thus, demographic heterogeneity (origin, farming experience, age), (im)migration or the increase of market-oriented individualism might differently influence various groups of farmers, guiding the behaviour towards co-operation and collective farming activities. When informal communication co-exists with the indispensable presence of state cadres also involved in everyday encounters, it has supplemented the state bureaucratic management system, modifying state initiatives to achieve an acceptable effectiveness in farming activities.

Inter-adaptation between farmers and the state

At the most local level of practice, state-designed and non-state or local informal arrangements engage in co-existence, supplementation and negotiation. With mutual interests in good harvests and the state recognition's of the impossibility of applying the command-and-control strategy, state cadres have adjusted and adapted to local arrangements while maintaining its effort at mobilisation. Farmers have also adapted, acknowledging the benefits of science and accepting the state presence and policies. This mutual adaptation is manifested in the consideration of the state sowing schedule and other agricultural policies, application of local planning to state planning for rice cultivation and the acceptance and sometimes even reliance on the state management structure.

Hamlet cadres report to the commune every day during pumping and sowing because of frequent changes in decisions about pumping activities at the commune level and in monitoring the implementation of the sowing schedule. Reports are made by phone to either the extensionist or the irrigation and rural road cadre. This practice lets the commune report its situation to the district DARD. For instance, reporting of seed lost can lead to support from the commune. Through regular reporting, cadres from hamlet People's Board describe the farming situation and explain whether and why the groups under their responsibility could not follow the state's sowing schedule, thus justifying the cadres' actions. Thus, reporting also allows local officials to give feedback, aiming for adjustment and adaptation both immediately and in the following season, and to express the upper-level state's un-official acceptance of local, on-the-spot adjustments. However, the current reporting system ceases with informing how farmers and hamlet and commune cadres have actively adjusted to the local situation and does not extend to seeking changes in mandates from the upper-level state. As far as most local cadres strenuously attempt to avoid criticism for not fulfilling a mandate (sometimes manufacturing false statistics), the Vietnamese system still lacks the effective feedback system that marks the *dialogue* or *responsive state*.

Uncertainties: The current foundation for changes

Systemic vulnerabilities and the need for changes stem from uncertainties throughout the field drainage process. In addition to natural and physical conditions, diverse human behaviours also give rise to much uncertainty. In the researched area, the existing arrangements seek alternatives because of persistent, unsolved problems: local cadres' identity struggle, the dilemmas facing group and zone leaders, individual cost–benefit calculations and the struggle for authority.

Local cadres' identity struggle

State cadres are at once the recipients of a mandate from the superior state, part of a pumping zone implementing local practices suitable that specific zone and a household member pursuing the household benefits. Local cadres undergo a continual struggle of identity — who they are, with whom and which group they shall comply — and become the focal point for flexibility.

The flexibility at the local level is driven not only by the local characteristic of flexibility ('*the king's rule stops at the village gate*') but also by changes in state characteristics. In the contemporary period, the sowing schedule and agricultural policies and their implementation generally have a softer nature and implementation. A local cadre narrates a story of the sowing schedule:

> Often, the office setting the sowing schedule has experience [how to scientifically track the hoppers migration pattern and farming tradition]. However, the schedule can never be 100% sure, so it is always followed by a sentence like 'depending on emerging situation in reality'. … Each district, using the city's schedule as a reference, will construct the schedule which is suitable for their district but must ensure the concentration of the sowing in one area, not like the leopard's skin. … The schedule has the nature of both encouragement and an obligation. It is sure that the farmers decide what to do, but if one field is sown against the schedule, resulting in damage due to brown hopper migration, that field with brown hoppers, which can be then infected by viruses, will probably cause an epidemic of rice grassy stunt and rice ragged stunt diseases for the whole area. … Therefore, the state has to transfer scientific information down to the people to farm, to protect the production of the locality. The duty of the agriculture office is very important in terms of making propaganda and mobilising farmers through the local state system to get information out to the ears of the farmers.... The mobilisation process also reveals the monitoring capacity of the state in each locality. (interview 08.04.2012)

Both an obligation and encouragement, production planning can become a real tragedy for both the superior state and local cadres. At the local level, the state tends to separate what is given in official mandates and on the ground talks. On one hand, official mandates affirm the obligation of policy implementation, which partly decides the local state's status in the management competition between communes and hamlets. Therefore, local cadres have been trained in the so-called '*communication arts*' in mobilising farmers to implement policies — the mobilisation task (*dân vận*). Skilled cadres strategically impose policy obligations so that farmers feel like it is encouragement. In this process, the state wants to prove that policy is made for the benefit of farmers, and the state only seeks to help achieve farming results. In this way, the state's emphasis on the encouragement nature of the policy is a strategy to maximise the results from local cadres' obligations. Generally, commune and hamlet cadres strongly feel the dilemma of policies when the policies need to be implemented.

However, the encouragement nature of policies in an obligatory state strategy also eliminates barriers, allowing local cadres to be flexible in implementation. The state cadres experience and understand that fluctuating water levels, unpredictable rain, differing dyke condition among and within zones and the possibility for the diverse patterns of social co-operation amongst farmers to be changed, delayed or scattered during the sowing time between zones and households within zone. In this case, the state practice of discursive power meets the exercise of farmers' discursive power, resulting in modifications of state management and implementation of policies.

Standing between the state and local farmers, local cadres are under various pressures. First, pressure comes from state sanctions and reprimands of those who fail to achieve production targets. Commune cadres are reprimanded if a great epidemic of brown hoppers or viruses occurs in their area of responsibility due to failure to implement the state sowing schedule. Failure in achieving the mandate for several seasons might lead to personal transfer of cadres as a form of sanction. The assessment of personal working capacity also takes into account the fluctuations of climate and challenges of local diversity (i.e. the dyke system). Second is the social pressure on cadres as prestigious neighbours. As mentioned, the prestige of the local cadres, most of whom are social leaders, is gained through good farming practices and high social spirit. Thus, local cadres have to maintain their status with their fellow farmers. As farmers and neighbours, local cadres understand and are part of local arrangements which sometimes contradict state policies and suggestions. Personal and communal interests can become pressures on cadres in the implementation of state mandates. Facing these potential dilemmas, local cadres must make local practices or state policy implementation flexible.

In addition, the uncertainty in state support can undermine the legitimacy of the state built on its reputation for '*caring for the people*'. Uncertainty in mandates gives local cadres additional reason for being flexible in policy implementation. As part of the process of the building the state's status, it tracks supports for policies, ranging from seed supply to compensation for seed or harvest losses because of flood. However, uncertainty, limited support and bureaucratic procedures[75] keep the supports from achieving the state's goal to be known for '*caring for the people*'. As stated at a pumping meeting:

> Farming is the farmers ourselves working together. I don't think even if we sow at the period of state schedule, if there is any damage, is there any đồng [VND] as support? I have 5 công, estimated to need 15 litres fuel for pumping. But water at the moment is still so high, how I can pump? And probably there is no fuel support. (Farmer, pumping meeting 16.11.2011)

Indeed, the late, non-transparent arrival and insignificant amount of state funds have created distrust and decreased farmers' expectations (interview 31.11.2011, 16.11.2011). Low expectations contribute to the state's decreased discursive power to convince households to take on risks or solve problems requiring co-operation.

In short, the local cadres have acted as intermediaries at the interface of practice negotiation. This research calls for re-consideration of the state as a single entity. Instead, the state structure consists of various groups of agencies, in this research, policy makers from the superior state and policy implementers or policy–practice intermediaries at the local state level. This research demonstrates that experience gives cadres more power and know-how to manipulate the state's mandate. The local cadres are, indeed, the Mekong Delta's form of *canaleros* (the water guards in the Autlan-El Grullo irrigation scheme in Mexico who '*are responsible for the translation of the technical guidelines and administrative orders from above, which they adapted to meet the varying needs, constraints and pressure generated at farm, field and block levels*' (Long 2001:74).

Problems of being zone leader

Group or zone leaders are both local social leaders and the lowest-level personnel in the state hierarchy. However, without any existing system of payment for service provision — management service in organising the communal works — the local leadership structure in the Party–state-managed

[75] In Thoi Lai, the high-quality seed support in the 2011–2012 main rice season was inadequate. In another case, the state promised to support the fuel costs during pumping, seed losses from flood damage during sowing and harvest losses due to pest. However, interviews and group discussions in the research area reveal that the support is small and, often, only comes in the following year. In a visit to Xuan Thang commune on 18.10.2011, the extension officer was very busy making contacts and discussing with the hamlet chiefs how to receive state support for production damage in 2010. The money was made available in 2011 after a list of field damages was sent to the district and then transferred to the province 1 year later.

and market-influenced Mekong Delta continues to rely on individuals' goodwill for its survival. As society gradually moves to a market orientation, this dependency introduces challenges to the effectiveness of local leadership as farmers have fewer resources and a smaller foundation on which to act for the sake of goodwill.

Group leaders, along with assistant(s), are in charge of organising all pumping-related works. They act under the state bureaucratic management structure and other local arrangements. Many elected during pumping meeting are hamlet leaders. These leaders are often named as co-operative groups, although many of these groups exist only on paper only (see Chapter 4 for details). Their responsibilities includes ensuring that all individual pumps are repaired and ready for the pumping date, calculating fuel sharing and informing farmers of and organising responses to delays in the pumping date (i.e. due to high water levels). Also, group leaders proactively handle dyke repairs and other crises. During pumping, they are in charge of monitoring duties. Often, they take regular walks to check for free-riding actions, detect broken dyke sections and mobilise others to fix them. Group leaders also receive and deliver fuel between pumps. The efficiency of work, whether done as planned or not, depends on the enthusiasm of each leader (given a negligible allowance) and the attitude towards co-operation of other farmers in the zone.

Willingness to pay for fuel, to join pumping and help in dyke repairs — all of these things cannot be strictly regulated and so depend on the goodwill of individuals, activated by encouraging state policies. Goodwill is the desire to do good for the community, sometimes at personal cost. Those acting with goodwill are regarded as having high social power. Many are local leaders and cadres, confirming the co-adaptation between state and local leadership structures. In other words, the state recognises and makes use of local leadership arrangements in order to allow maximum opportunities for effective policy implementation.

Individuals' goodwill is significantly influenced by their past experience and present conditions. Goodwill depends on whether fellow farmers appreciate when others demonstrate high social spirit in working for the community. Appropriate responses to this work are agreeing on farming co-operation and showing appreciation and respect for the work of organising and monitoring the pumping process. Farmers themselves suggest contributions of 1,000–2,000 VND (0.05–0.1 USD) per *cong* as an allowance for the group leader. However, not all farmers give, and not all group leaders want to receive the allowance. One group leader opposes receiving that compensation:

> Let's do as before. I don't receive money but am responsible. It is easier to communicate to each other [he and other farmers]. If I receive anything from you, even VND 500, it is taking money from you. I am working as group leader because I want it and for the community' (21.12.2011)

State appreciation through giving status (i.e. position as a cadre) and financial and in-kind support (i.e. getting a priority plot in technology training and credit programmes, interview 29.11.2011) also offer motivation to act on goodwill. However, uncertainty in farming co-operation leaves the community and the state no other option but to rely on the goodwill of some individuals, but over time, that uncertainty reduces motivations to act with high social spirit. Today, goodwill is based not only on the Delta ideology of village love and neighbourly relations but also on individual tolerance of individualism and free-riding actions, which result from a market orientation. Living expenses are increasing, and the sense of community is eroding because of migration forces and the tendency of households to devote their available resources to the household's benefit.

So far, voluntary contributions to finance and labour for the community's good have survived, inspired by the remaining social spirit and persistent encouragement from hamlet cadres. In the researched area, the negligible return for the extra work, the uncertainty of its effectiveness and the difficulties in gaining co-operation make the farmers hesitate to take on leadership responsibilities. As one participant states, '*someone else should do it* [become the group leader]. *I can't do it. I have a lot of work* [in the household] *to do*' (meeting 29.11.2011). However, there are still zone leaders because of attitudes such as the following:

> No one wanted to do it. Seeing that, I thought I would just do it as sowing early has a lot of benefits, such as high yield. In fact, I have a lot of other work to do, such as taking care of my own field. Others always calculate [for their own benefit], so they don't want to do it [being leader to organise the pumping]. (interview a group leader, 29.12.2011)

Goodwill is not yet missing in the community but is increasingly challenged by continuing changes in market-influenced society.

Given these challenges, farmers seek non-voluntary alternatives for conducting collective farming activities. Field drainage is contracted to individuals or companies with high-capacity pumps. Farmers joining the contracted pumping only pay an amount of money for the total service, from repairing the dykes to pumping until individual ridges are seen. This method requires individual payments based on land holding; thus, it results in fairer contributions and helps reduce the burden on zone leaders. On one hand, the step towards more accurate calculations lessens pressure on the system which has relied on goodwill challenged by the market mentality. On the other hand, monetary equality results in fewer complaints about production and reduces feelings that one farmer is doing more than others. It can undermine the social traditions of '*village love and neighbourly relations*' as farmers have fewer occasions to sit together, discuss and work collectively.

Individual cost–benefit calculation

The fear of higher costs due to uncertain water level and dyke condition is the main barrier to co-operation. An interviewee expresses what households fear:

> 'It is raining quite often, and after every rain, water is getting really high. First, we can't afford too much fuel [during the pumping process, if the water gets high, farmers simply have to continue pumping and put in more fuel]. Second, when the water is still high, if we sow, if something happens, it is very expensive to buy new seeds; it is even impossible to buy seeds at this time. Our income is based on rice, so if something happens, we could not afford our living for at least 3 months. (interview 30.11.2011)

In rice-farming communities, some households are more vulnerable than others, especially small landholders from whom rice cultivation is their main income source. These households have less appetite for risk-taking than other farmers.

With the costly modern life and large high financial differences between households, individual costs and benefits are becoming more critical in households' consideration of joining collective actions. However, the old arrangements still in place create an environment with a mixture of local arrangements. Consequently, individuals' actions differ. For instance, in the planning stage for fuel sharing, participants selected the counting-each-coin strategy in a signed agreement as no farmers could cover the others' costs (meeting 24.11.2011, 29.11.2011). However, the farmers choose to act through a more diverse system of sharing, rather than accurate calculation and paying.

In some cases, paying neighbours can seem a sensitive issue to some farmers, especially households with less income. The economic capacity of these poorer households directly affects their tolerance of free-riding, sometimes causing conflicts. To those farmers, payments that create fairness amongst farming households are not '*weird*' at all. The current institutions are neither similar to the established arrangements nor completely new. They consist of some established features (the feeling of awkwardness at monetising relations) and more situated ones (the necessity of sharing fuel costs or, at least, discussing and partly implementing it).

Cultivation planning for 3 seasons brings farmers into competition with their neighbours, which the state can strategically exploit. A hamlet cadre confidently states that:

> It's not a problem [that many farmers did not join the meeting and hesitated to agree on the date]. When people see and hear the sound đùng đùng of the starting diesel pumps, they will be anxious and will join us. (after meeting 12.11.2011)

Driven by competition, some farmers, who hesitated to join the agreed-upon pumping event, rethink their decision when seeing others start the season. At that moment, they might decide to ignore or accept the risks of join pumping with their fellows. In this situation, the economic factors become both constraints and opportunities for co-operation. Each individual is bounded into a competitive game of farming but also considers collective action as a potential way to achieve better outcomes.

Decision-making process and individual behaviour are influenced by state strategies and individuals' social prestige and ties, previous experience and enthusiasm, which can marginalise important but uncertain factors, such as water level and climate. This research proves that (1) the open structure of rural Mekong Delta communities offers less space for social ties and pressure on individuals, except in the case of kinship ties; and (2) official sanctions and local rules guiding co-operation are impossible as the state mobilises but does not command local activities and desires to promote the image of grassroots democracy.

Struggle for authority in negotiations

> The hamlet People's Board have to be here to check [the co-operation in pumping]. The group leader and other farmers are equal [in authority], so it is difficult to tell the other to follow certain things. Who will listen? It is difficult! (Group leader, interview 24.11.2011)
>
> Don't dare to! Because we have neither the responsibility nor the authority to force anything. We are group leaders to organise the water pumping; other than that, we have no power. (group leader, interview 29.01.2011)
>
> - Common responses regarding group leaders' lack of authority to resolve conflicts – 2011–2012 field drainage season

The field drainage organisation before sowing for the main winter–spring rice season is not based on simply voluntary co-operation for the communal benefits of all. Some farmers have more benefits than others, creating potential for poor coordination and free-riding actions. However, sanctions are not enforced through authority but, rather, through social pressure, mobilising, convincing and negotiating.

Negotiation happens amongst farmers over the date, forms of co-operation, sharing fuel and persuasion others to follow the agreed-upon plan or co-operate fairly. However, farmers do not have authority over each other, and farming is a matter of individual interests, one cannot '*go strong*' (or using strong words and ordering compliance) on violators. Zone leaders face the same problem, even though they serve as managers and mediators in case of conflict. A group leader describes his action in a conflict situation:

> Because we are all farmers, we can only discuss with each other. If agreement can't be reached, we have to hand the case over to the state [hamlet People's Board]. We are farmers. We have no authority. We need to request support from the state. (interview 29.12.2011)

Thus, individuals retain the power of freedom in farming. This ideology of freedom contributes to fairness in local negotiations of farming co-operation but becomes a constraint in conflict situations.

Then, insiders must search for external authority, strengthening the persistent presence and function of the state. The state mentioned in the preceding quotations is hamlet cadres. Farmers consider these state assistants who receive an allowance, instead of salary, to be part of the state. However, state authority does not ensure a complete solution either. Despite farmers' expectation that the state will help overcome negotiating failures, the effectiveness of hamlet state involvement depends on the prestige of individual cadres, the strategies applied, the farmers and the specific case.

Hamlet cadres can provide some assistance when the typical strategies of encouraging, negotiating and mobilising through logistics and group-signed agreements fail to convince difficult households. One method is *intensive mobilisation*, or making regular visits to difficult household to convince them to join the group activity. This method reportedly works when the household starts feeling ashamed before the hamlet People's Board and other farmers. In addition, CPC leaders can report difficult cases in paper to the commune and publicly criticise households at meetings. However, this extreme kind of shaming rarely happens because it '*is complicated and creates further conflict between the neighbours* [the violator and the reporter]' (group leader, interview 29.12.2011). Thus, the relative decline of the state's command-and-control power is accompanied by cadres' difficulties in achieving compliance at the local level. In a context of growing individualism, it is increasingly common to observe struggles or outright failures in negotiation.

Today, local cadres are trained not to force farmers in ways which might create a bad image for the state. A farmer states that, '*sometimes, it might be better to be like before* [1980s] *when the state just ordered the people what to do*' (interview 14.12.2011). This statement does not favour the strong state control to which farmers objected during the collectivisation period but expresses a wish for a mechanism to resolve conflicts. The Southern Delta's inhabitants retain their interests in individual farming which the state has been officially recognised since Doi Moi but are seeking effective mechanisms, in addition to the remaining value of kinship and social ties, for organising various collective works required for market-influenced production.

Conclusions: Negotiating field drainage at the interface

Field drainage entails organising the collection of diesel pumps to pump water from fields in order to enable early sowing wherever gravity drainage is not feasible. Under the influence of natural and physical conditions and state interventions in and regulation of rice production, field drainage before sowing in the winter–spring rice season presents a space for social networking, institutional rehearsal and negotiation among various actors, including the state, local cadres and farmers. During this process, policies are reviewed, adjusted and adapted to create institutions fit specifically for a group with certain common interests in a particular activity at a certain point in time. Stakeholders join negotiating field drainage, motivated by various interests that might overlap, supplement or contradict each other.

Within field drainage, practice is negotiated by various stakeholders, including local cadres, farmers and pump owners (in the case of contracted pumping). A number of factors are at play: natural and physical conditions, social institutions and state incentives offered in policies and mandates. Over the history of state intervention, stakeholders have not only accepted the state's presence but have also grown reliant on state support, especially amid the increasing difficulties in social co-operation in the Mekong Delta caused by the rising spirit of individualism and capitalism. Figure 5-14 visualises the process from sowing schedule formation through local interactions. The expectations for and the nexus of relations clustered around water control for the very important winter–spring rice season best illustrate the nature of state–society interactions in the contemporary Mekong Delta.

Figure 5-14: From policy to implementation: The state sowing schedule in the case of Can Tho (made by author)

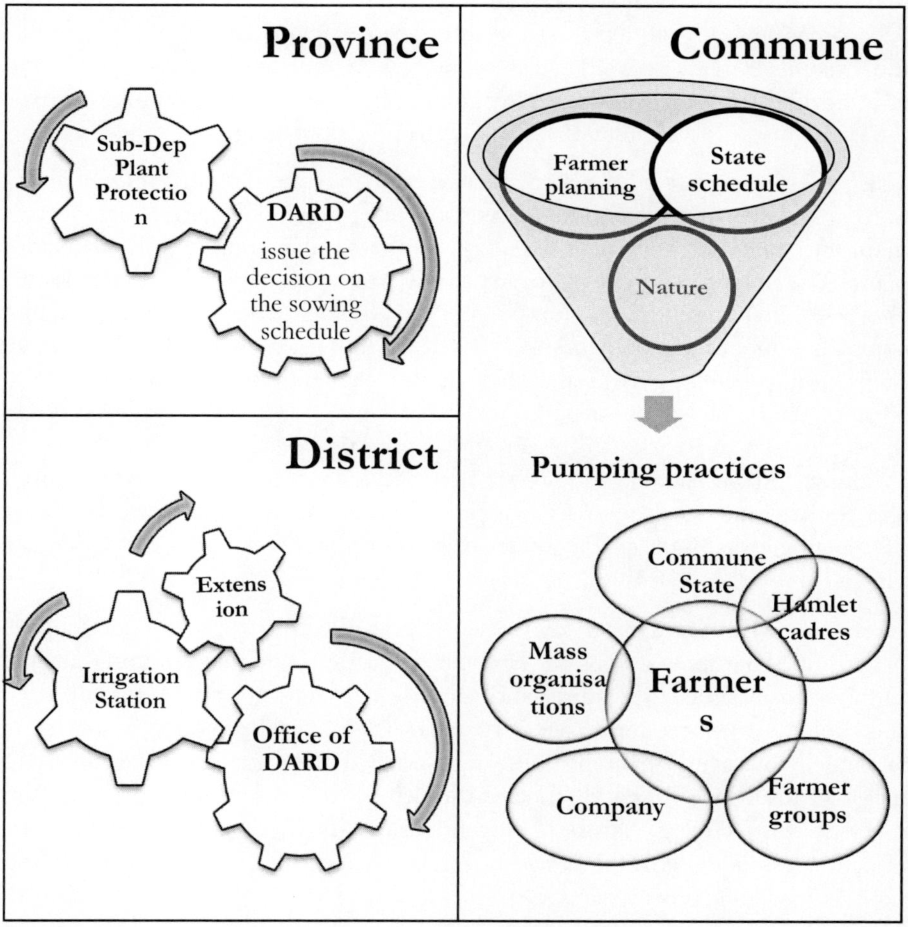

The state is present in local farming as the manager and facilitator of production process, or in other words, the implementation of state policies. Most state agricultural plans and programmes are supported by scientific evidences, and their implementation follows a rigid, standardised, top-down procedure, from province to district to then commune. For instance, the sowing schedule, which the state bases on local farming experiences (e.g. sowing at low water in October of the lunar calendar) and scientific knowledge (e.g. avoiding the plant hoppers), has gradually become part of local farming planning. Until reaching the commune level, production planning is transferred through official meetings at the central DARD office, district DARD offices and CPC meeting halls as the superior state disseminates information to at least guarantee fulfilment of the production targets. However, the meetings are merely the start of interactions and negotiations among state cadres, farmers and sometimes private actors at the interface of field drainage process. Whilst the existing bureaucratic checklist of tasks and meetings remains rigid, with little flexibility in outcomes, diverse natural, physical and social conditions make the implementation of state plan difficult to predict or judge.

Thus, the direct and informal interactions and negotiations in the hamlet and pumping zone, in which the state is only one stakeholder, are more decisive than the state bureaucratic procedure in the implementation of production plans. The practice of field drainage activity results from various arrangements, choices of action and interaction process which involve agreement, negotiation, contradiction and supplementation. Based on this research, the hamlet meeting or pumping meeting marks the start of the planning for field drainage in all areas of the hamlet, with or without the presence of a farmers' group and with the facilitation of hamlet cadres throughout the process from pumping meeting to sowing. At this stage, varied natural, physical (e.g. water level, dyke condition) and social conditions (e.g. social organisation) construct the diverse landscape of meeting organisations, negotiation methods and results of field drainage practice.

Interaction and negotiation take place during the 3 stages of field drainage: preparation, pumping and costs. In the preparation stage, the pumping meeting fulfils bureaucratic procedure and provides a starting point for different stakeholders to enter the process of negotiation. Field drainage practices are facilitated by state planning but are more dependent on local, everyday negotiations in public spaces, such as the rice fields, markets and cafeteria. With diversity influencing farmers' incentives and social arrangements, the effectiveness of state mobilisation is only a piece of the puzzle that determines field drainage and sowing activities.

Starting with meetings, this study illustrates a pattern of negotiation in which individualistic farmers have the freedom to discuss the timing of pumping and sharing of fuel for collective field drainage in zones. However, the freedom to speak in meetings does not necessarily mean that decisions and negotiations are based strictly on a system of fairness. Rather, there operate power relations in negotiation depending on social prestige (mainly gained by age and good farming practice) and gained through strategic, physical conditions. Whether to raise an opinion or not, to discuss an issue immediately or later, to object to an agreement or stay silent, actions during meetings result from complex considerations of bureaucratic, economic and social factors and individuals' socio-economic status and power. During the negotiation process, the private nature of farming activity means that the power and right to speak depend on social prestige based on age and farming capability and less on landholding or gender.

Zoning practices present the first uncertainty and alsothe foundation for change. Zoning for field drainage is subjected to change based on weather and farmers' commitment to participate in field drainage. Zones are both a convenient bureaucratic and systematic way to connect residents in a forum for social co-operation and state involvement and a reflection of the economic interdependence which has played an important role in social arrangements of the Mekong Delta. Throughout the process of field drainage, natural (e.g. water level) and physical conditions (e.g. dyke situation), which, by nature, are uncertain and diverse, provide both constraints and opportunities for local farming co-operation and practice.

Fuel sharing in field drainage illustrates the diversity in arrangements which results from interactions between older social arrangements based on mutual sharing and newer monetary calculations of cost. The selection of a more accurate sharing method arises from the pressure of persistent unfairness in sharing under older methods and the increasing financial and social mobility that challenge the social system of continuous, in-kind community sharing. During the pumping event, at the field house which serves as a public space for neighbours to gather for eating and drinking immediately before, during and after during the season, the individual style of living and farming has become less dominant in in this market-influenced era, leaving space for institutions that promote mutual help and sharing among farming neighbours.

Field drainage also reflects the intra-household relations in farming families. This research confirms the gendered task division in Vietnamese society; men are assigned heavier works, while women are responsible for lighter tasks, including weeding, seeding and domestic chores. Particularly in field drainage, the intra-household relation illustrates the complex process of decision making which does not necessarily align to the public representation of family members. The task division in farming in the contemporary Delta reflects internal negotiations based on trust giving and harmony building (e.g. women strategically give men more space in public).

The results of hamlet or zone planning for field drainage depend on the negotiation of rules and norms and the bargaining of individual behaviour. Behaviour varies among individuals based on the social negotiation of (1) individualism, in which households attend to their own costs and benefits in deciding whether co-operation is a good choice in certain circumstances; (2)

state efforts to continuously mobilise collective activities; (3) social pressure to be good neighbours, though this is less influential than other social characteristics (i.e. competitiveness among farmers); (4) sentiments based on past experience (e.g. negative or positive sentiments towards a person, agency or way of organisation); (5) the ideology of living at peace with neighbours and the high social spirit to do good for society; and (6) the tolerance of free-riding actions by some individuals. Individual decisions follow emotional, moral, social and economic rationales; of these, economic rationales are becoming more dominant.

Decisions about collective works are made by individuals who hold private land and have private incomes which give them the power to negotiate and fight for their own benefits. Individual resources and emotions mean that the extent of influence from state ideologies and natural and physical conditions differ among individuals and activities. Among the factors that need to be considered in field drainage, natural and physical conditions clearly trump imperatives from the state sowing schedule or other bureaucratic mandates. Economic logic and interdependence play dominant roles in decisions on co-operation. Individuals engage in competitive farming but also regard collective action as a potential opportunity to achieve better outcomes. Additionally, other moral ideologies, such as a high social spirit, co-exist with individualistic lifestyles in the open-market, rural Delta, creating a more diverse system of sharing pumping duties and fuel costs than accurate calculation and paying.

Thus, in field drainage, the most decisive influence on co-operation is the ecological conditions which lead to economic dependency between farmers in the same zone. When co-operation is a clear solution to a common threat, risks and uncertainties might be neglected. In addition, other factors, such as social norms and state influence, affect individual co-operation behaviours. However, the open structure of rural Mekong Delta community offers less space for social ties and pressure on individuals, except in the context of kinship. Official sanctions or local rules guiding co-operation are still far from possible as the state mobilises but does not command local activities. With influences from many factors and uncertainties about the costs and benefits of collective action, co-operation in field drainage varies among areas. Differences are recorded in the selection of calculation methods for fuel sharing, the implementation of meeting agreements, the fairness of labour and fuel contributions and the intensity of conflicts.

Among the many contributions to collective field drainage, free-riding actions appear to impose burdens on others, particularly on the hamlet cadres who have a responsible to keep state and local plans moving forward, on enthusiastic farmers who have a high social spirit and on the most vulnerable farmers who have fields in which it is impossible to start farming early without collective pumping. Therefore, an activity such as dyke repairs lies at the intersection of economic vulnerability and interdependence, of state encouragement and voluntary spirit. With the lack of social authority — no one, not even the zone leader, has more power in negotiating private farming practices — constrain local arrangements of co-operation. In the current situation, field drainage depends on the tolerance and goodwill of individuals who perform the work of others. In general, social arrangements in the private farming system are regulated by local institutions shaped by the local structure of prestige and economic-based cost–benefit calculation. When obvious and perceived by all, economic interdependence and mutual vulnerability create great incentive for co-operation and collective action. The leftover social engineering (especially for cadres) from before Doi Moi and the continuous ideology of cadres as models help overcome the communal hurdles increasingly posed by capitalism.

Field drainage activities are organised at the interface of the state bureaucratic structure and the local negotiation of practice. Prominent at this interface is the complexity which creates local dynamics of multi-identities, best illustrated by local commune cadres and hamlet state assistants. Close to and understanding practice and reality, local cadres adjust the state schedule so that it is realistic and unofficially recognise the value of arrangements or institutions established locally. In the bargaining of individual behaviour, local cadres act differently depending on their years of experience as state cadres (e.g. young cadres tend to act more rigidly, while more experienced ones are more flexible and adapt state mandates and locally suitable arrangements) and the natural, physical and social situations. Especially when state mandates contradict local conditions, policy tends to be disseminated at the local level as encouragement. While deciding with whom and which group to comply, local cadres become the focal point of flexibility. The state's attempts to retreat from financial and management duties while maintaining a minimal role in the sector have contributed to uncertainty in mandated obligations and state support of farming. This situation has driven local cadres and farmers to be flexible in policy implementation, while low expectations have reduced the state's discursive power to convince households to take risk or engage in co-operation.

Amongst the actors involved, the zone leader in field drainage stands at the centre of interactions between the state bureaucratic structure and local co-operative arrangements. Many zone leaders are also local leaders or cadres, demonstrating the co-adaptation between state and local structures of leadership. The efficiency of work and whether it is done as planned depend on leaders' enthusiasm and farmers' attitude towards co-operation with others in the zone. Local cadres' goodwill is significantly influenced by past experiences and present conditions (e.g. whether farmers respond appropriately to cadres' efforts to work for the community with a high social spirit). The cases reported in this research prove that goodwill is not yet missing in farming communities, despite the growth of individualistic tendencies in rural areas as a result of expensive, busier lifestyles and free-riding actions. However, voluntary work is challenged by the continuing changes in the market-influenced life. This situation has prompted the search for monetarily fair alternatives, which might undermine the social traditions of '*village love and neighbourly relations*', giving farmers fewer occasions to sit together, discuss and work collectively.

Field drainage organisation by collective actions still dominates the researched area, driven by farmers' economic dependency. Tolerant individuals with high social spirit continue to compensate for others' free-riding actions or unwillingness to contribution. However, this tolerance based on high social spirit and disregard of inequality makes current arrangements fragile. The current system based on social cohesion and provider–consumer service offers no mechanism to sustain arrangements. Thus, participants have sought alternatives and have started to use one in a trial stage: contracted field drainage. While vulnerable, the state's status is reinforced as it continues to be a social leader and manager, providing certain authority in overcoming the barriers to co-operation. Consequently, arrangements are in constant negotiation and can create constraints or opportunities for various stakeholders. Those constraints and opportunities create the foundation for changes or the making of new institutions.

Studying field drainage activity enables capturing the inter-adaptation of the state and farmers. Electrical pumping stations and collective field drainage can be seen as the interfaces in which the knowledge and capital to acquire and use technology guide its implementation. The adoption of collective and individual practices demonstrates the heterogeneous reality of farming practice in the area, with many options for adaptation during the process of learning by doing. Farmers' practices and local arrangements influence state implementation of policy and, in turn, are influenced by and adapted to the state bureaucratic procedure for production.

The overlapping of state cadres and local prestigious individuals and social leaders again proves the adaptation of the state as farmers and local cadres rehearse the state policy of collective farming and make it suitable for intensive, commercialised agriculture production. The state recognition of a 2-step negotiation process implies the acceptance of a bureaucratic state which permits flexible implementation. In the present system, adaptation through trial and error takes a long time when information about failures and changes are officially transferred to the policy-makers only at the end of the season. The negotiation and adaptation process illustrates the everyday dialogue between the state and the involved stakeholders that illustrates the dialogue facet of the Vietnamese state, as affirmed by Kerkvliet 2001b, Kerkvliet 2003).

Field Canal Dredging

Despites difficulties in achieving co-operation among farmers, working collectively for fast, effective drainage is dominant in the researched area. Given farmers' economic dependence, collective action in pumping offers a solution to the threat of too much water for early sowing. However, where a common threat and economic dependency are less visible, the value of collective action decreases or even disappears. Field canal dredging activities generally are not urgent and so support individual solutions and make collective ones time and cost intensive, decreasing the incentive for collective activity. In canal drainage, the difficulties of the costly modern lifestyle, labour shortages and individualistic farming become more prominent. The impacts of poor canal maintenance are indirect and different for each household, and the availability of even uncertain state support reduces the chances for collective action in dredging the selected canals.

The following sections analyse negotiations of field canal dredging projects, in which the district state is the main investor, the local state is the field assistant, and farmers are the payers. The analysis shows how, in the changing context, farmers have moved from being beneficiaries, payers and negotiators to being dependent on the state budget for dredging field canals (see Chapter 4 for the state classification of canal types and budget sources). The analysis of field canal dredging suggests that co-operation takes place under very sensitive conditions in an increasingly capitalistic and individualistic setting. Therefore, the state attempts to create those conditions for its retreat need to be revaluated in order to ensure the effectiveness of dredging effectiveness, state control of production and the state's reputation for caring for the people.

Canal dredging is vital but neither for all nor at all times

The field canals in Can Tho are approximately 5–7-m wide at the end of hydraulic system immediately before entering the fields. At the start of hydraulic irrigated agriculture, farmers manually dug field canals of 3–4 m to convey water from larger canals to individual fields. Given bank erosion and sedimentation from the silted water, the canals were dredged manually with soil buckets (Picture 5-14) in order to convey irrigated water and enable transport by *xuồng ba lá* or *tamban* (the 3-piece boat).[76]

[76] Each farming household had a soil bucket to use for dredging and bringing soil to reinforce the local dyke protecting a house or rice field.

Picture 5-14: A bigger form of *tam ban* (Huynh Linh, 2011); water bucket – a similar form without a bottom is used as a soil bucket in the Mekong Delta (VSH, accessed 2014[77]).

77 http://www.xomco.vn/dien-dan/xom-ky-vat/doi-thung-ganh-nuoc-cua-noi (accessed 17 April 2014)

As mentioned in Chapter 4, dredging is becoming necessary to restore or widen canals. Sediment can be removed by dredging manually with soil buckets or mechanically with small or big dredges, which is the most commonly used method. While big dredges dominate large canal projects, small ones remain popular for their convenience in moving among small field canals. In general, soil buckets are still used for dyke reinforcement, and canal dredging is done by dredge machines. The current mechanisation in canal dredging is the joint result of the booming of private business in agriculture and construction and the need for a solution to the labour shortage created by the out-migration of young, rural inhabitants to work in the industry and service sectors in cities.[78]

Within the state hierarchical system, field canal dredging falls under the responsibility of the people or farmers. The usage of machines transforms the activity from labour to money dependent. Farmers organise and pay for the dredging of field canals, while responsibility for source canals belongs to a decentralised state structure of relevant agencies and an inconsistent system of budget allotment (see Chapter 4). Giving farmers the responsibility to organise, pay for and control field canal dredging is amongst current state efforts to decentralise and share the financial burdens of the state budget. Although assigned to the people, field canal dredging remains a local affair which the state believes it needs to monitor and state control. Consequently, commune and hamlet cadres have duties to mobilise co-operation in dredging and, along with residents or the local state, find solutions for persistent problems, such as conflict situations and a lack of money.

Technically, the conditions of the Mekong Delta require dredging field canals every 3–5 years (interview 04.07.2011). However, decisions to dredge canals vary by time and individuals, depending on natural hydrological requirements, financial capacity and social views of the urgency of the need. Research in Can Tho reveals that the later factor is the main driver in discussions, decisions and contributions to dredging field canals. Dredging a canal is only urgent when that canal cannot serve its functions, which occurs only at specific times: harvest, when the canal has to be big enough to carry large boats to transporting rice, and the dry season, when shallow canals provide almost no water for the fields and no space for boat transportation. Only then can the

[78] In the research sites, all families report changes in rural inhabitants' career orientation which have led to a lack of human resources for agriculture. The Office for Social Evaluation and Consultancy - OSEC (2011:7) reports a significant decline in the agricultural workforce and increase in the industry, construction and services workforces.

state mobilise the participation of residents in dredging. A hamlet cadre states, '*it is so tiring. We* [the hamlet People's Board] *couldn't do anything, … just wait until the canal gets stuck, and no one can use it. Then we can ask for dredging*' (hamlet cadre since 1980s, interview 30.11.2011). Consequently, most dredging projects take place in the dry season (February–May) when the canals are very shallow or even dry up, revealing to farmers the imperative of dredging. In contrast, during the flooding season, as in 2011, '*the farmers have no interest in dredging the canals*' (district cadre, interview 04.07.2011). Social mobilisation and collective action are taken only when the threat is obvious and perceived by farmers.

However, even when the threat is clear, not everyone perceives the same threat potential and feels the same sense of urgency, posing difficulties to carrying out dredging projects. In the open, dense network of canal system, the dependence of fields on a single canal varies. Some fields have access to more than one canal, which makes the owners of that field less sensitive to the shallowness of a canal and thus less motivated to dredge it.

Although all the parties involved generally feel the urgency to dredge a shallow canals, various factors affect the motivation to contribute monetarily. For instance, a hamlet cadre describes the situation of canal 500 (hamlet Truong Thuan, commune Truong Xuan B), constructed in the 1980s and dredged in 2010:

> The canal is about to be shallow again because it couldn't flow to the UC canal. This piece of land now belongs to another commune, so it is difficult to just dig the canal through it. The canal is so shallow that boats [ghe – bigger than tamban, the smallest kind of boat made from 3 pieces of wood] couldn't come to the canal. I don't know what to do. Only some farmers hire the small dredge to dredge the canal section in front of their field so that they can withdraw water. But it is impossible to make it for the whole canal because the people don't agree. I said it would be just 6,000 one scoop and about 4 scoops at one place, and so on. It is nothing, but they don't agree. (hamlet cadre, field observation 14.02.2012)

Paying for dredging can be a solution to labour shortages but faces the obstacle of farmers' limited financial resources. Household financial capacity is a decisive factor in canal dredging. In this case, the amount of hundreds of thousands VND (around US$10–20) might be acceptable to some farmers but financially disadvantageous to others. Adding to the rationales for dredging in the dry season, dredging is easier when canals have less water and is more financially feasible after farmers have profited from the harvest of the main, and most successful, rice season.

The fairness of the traditional cost-sharing method has been undermined by costly modern life. Amid inequality, the method becomes a constraint to co-operation by farmers in dredging. In the researched areas, the payment arrangement for canal dredging was based on local units of *tầm* (equal to 3 m). Farmers pay for the number of *tầm* of their fields adjacent to the dredged canal. The cost for 1 *tầm* is calculated based on the number of soil cubes removed from the canal; each soil cube costs VND 6,000–8,000. The cost of 1 *tầm* is shared by the owners of the fields on either side of the canal. This calculation method has long been practiced in the area, although payments do not correlate with usage of the water or canal. Farmers report that no alternative has been used. If a farmer unluckily has even a small field that runs the length of the canal, the farmer must pay for many *tầm*. Such households often hesitate to join canal dredging (interviews 30.11.2011, 14.12.2011). Consequently, the established social arrangement is under pressure to achieve monetary equality in order to gain more participation from farmers. Recently, some local strategies for equal sharing have been applied; for instance, others partly share the large amount owed by the farmer with a field running alongside the canal field (interview 14.12.2011). Overall, field canal drainage faces timing and coordination issues, different than the reliably cyclical regime of sowing and field drainage. Additionally, the costs for dredging are perceived as extra, random and not everyday costs of rice farming. Consequently, achieving co-operation in dredging field canals is most problematic when increasing numbers of canals get too shallow to be used or permanently dry up.

Thus, the context of higher costs and possible alternatives has shifted relative fairness from priority in the consideration of effectiveness. Canal dredging is fully individualistic; thus, individual behaviour is decided by cost–benefit calculations and guided by economic factors, while the significance of other moral and social factors is lessened. Rational individuals decide co-operation behaviours (Ostrom 1990:6), which leads to the tragedy of the commons (Hardin 1968).

Organising field canal dredging and unsolved problems

Although dredging field canals seems to be a matter for farmers, the local state has the tasks of monitoring and ensuring that a certain number of kilometres of field canals are dredged every year. For instance, the target might be to dredge 1,000–2,000 m of field canals annually (interview 14.12.2011). Dredging, then, is an affair in which the hamlet state takes charge.

With technical support from the commune, hamlet People's Board members have the responsibility to organise, facilitate and arbitrate the dredging projects.

Different than the regular groups of field drainage and farming which might be supported by the established network of neighbours, the collection of farmers who have fields and houses along a canal and often belong to more than 1 farming group is led by the hamlet cadre. The tendency to allow a greater state presence has increased as canal dredging has become more dependent on state financial support. The state is becoming the 'go-to' party when problems arise, or in other words, the state has achieved a target of using discursive power: making its presence and assistance seem natural.

Different than the formal procedure to transfer the production plan and sowing schedule, the field canal dredging mandate assigns local cadres the general task to organise either meetings or household visits to discuss the dredging project and secure contributions, such as clearing trees on the canal banks and agreements to pay for the dredging. The local process of dredging a field canal is described in this section. When a canal is perceived as '*urgent enough to be dredged*' (many farmers have complained that it malfunctions, which the hamlet cadre has reported to the commune), the irrigation cadre surveys the canal's condition, length and width and estimates the cost of dredging. Often, a meeting of hamlet and irrigation cadres, commune leaders and farmers with land along the canal is held next. Depending on the agreement achieved in the meeting, a dredge is hired, dredging is started immediately or later, and money is either collected by 1 person (often a hamlet cadre) or paid directly to the dredge owner. For dredging, the hamlet state organises meetings similar to field drainage organisation but otherwise follows a very different procedure. Unlike field drainage, dredging does not have a strict deadline to be completed before the next low water cycle, so the meeting only takes place when the majority of the water users of that canal agree on the urgency. Therefore, to a certain extent, state involvement in canal dredging facilitates requests, and the meeting responds to the situation on the ground. Consequently, any agreement achieved at the meeting likely is realised in practice, although the problem of non-paying is common after dredging.

Canal dredging, despite its necessity to the people and the available facilitation of the state, still faces obstacles to participation in collective action, such as free-rider actions, leading to the current application of ad-hoc strategies. Two issues typically arise during field canal dredging: no or delayed payment for a completed project and objections, resulting in a dead-end canal.

No or delayed payment from farmers after dredging project is not rare, causing a decrease in local state interventions in the dredging process. For instance, in the 1990s, hamlet leader(s) representing a group of farmers signed a contract with a dredge owner. The cadres collected money and mobilised the payment process. However, they failed to ask all farmers to pay, causing debt, and gradually lost their motivation, leading to state agencies' responsibility to monitor the sector. A former cadre of hamlet Farmer's Union tells of his experience of serving a representative in field canal dredging. With disappointment, he states:

> Field canals, we give up. Before dredging, they talk about it so smoothly, but when it is dredged, they always talk the opposite [not agree and/or delay to pay]. … This canal, my household still owes the dredge owner more than VND 14 million. Dredged around in 2004, the people didn't pay. In that case, the dredge owner brought the case to court. I made a list of the households which didn't pay. However, the court said the amount of money for each household was not big enough to start a legal case. Some owed VND 300,000–400,000; some 500,000–600,000; the biggest amount is more than VND 1.2 million. As I am the representative, the total amount was enough [to start the case], but I was only the representative. I think now the dredge owner has given up, I haven't heard from him for a while. It is almost 6, 7 years. The canal is already shallow again. (interview 21.12.2011)

Tracking the reasons for changes in dredging mobilisation or, in other words, in co-operation in dredging shows the influence by moving away from collectivisation in the 1980s. When farming was returned to individual households, it eliminated inhabitants' obligation to contribute to communal activities, including canal dredging. In 2007, households were also released from community service (10 days for public work) under Ordinance 15/1999/PL-UBTVQH10, which ended the responsibility to carry out public work. Since then, canal dredging, either manually with soil buckets or fully mechanised, has been an individual matter. In other words, it appears to be regarded as extra burden on top of regular, accepted farming costs. In addition, dredging illustrates the almost complete freedom of farmers whose holdings are scattering over a large area.

The cases of no or delayed payment and objections to canal dredging confirm the tragedy of the commons (Hardin 1968) and the logic of collective action (Olson 1965), in which '*rational, self-interested individuals*' act in their own interests and, except in small groups, '*will not act to achieve their common or group interests*' (Olson 1965:2, cited in Ostrom 1990:6). According to Ostrom (1990) criticisms of this school of thought on human behaviour, some will chose to engage in free-riding actions so as to have less costs but equal benefits from canal dredging. However, the reported cases have not seen many free-riders, although economically, '*whenever one person cannot be excluded from the benefits that others provide, each person is motivated not to contribute to the joint effort, but to free-ride on the efforts of others*' (Ostrom 1990:6, summarising Hardin's and Olson's work). Thus, other factors must influence decisions, which are based on the benefits and costs of actions and their perceived linkage to outcomes (ibid:33). A hamlet cadre narrates the story of a dredging project in the 1990s:

> Back then, I calculated everything [the canal length and width and the estimated cubic meters of soil] for that canal. It was done fast. All farmers paid for their part, except one. That farmer denied paying since his field is located at the meeting point of 2 canals which always have enough water. He said that he will not use water from the projected canal. After the completion of the canal, some farmers along the canals denied letting that farmer to use the boat going through the canal. Conflicts happened, and hamlet and commune cadres were called to find a solution. After sometime, the farmer had to pay back his share to be able to use the canal. (canal dredging meeting 20.08.2011)

Thus, social pressure exerted power over free-riders. The group's fight for equality against the aggressive actions of some individuals forced the free rider to correct his behaviour. However, as Ostrom (1990:7) criticises the notion of the prisoners' dilemma in which they cannot change the constraints imposed on them by the district attorney, some users can change their constraints. For instance, those who have access to more water sources for irrigation and transportation felt no or little pressure from social sanctions when they engaged in free-rider actions. Without command-and-control state management, the coercive power of local institutions and rules constructed with and without the authority of (state-designed) institutions (Ostrom 1999) might have no influence in the open society of the Delta.

Local strategies and state involvement in the sphere of famers' responsibilities

Although securing agreement from all farmers and mobilising them to contribute to dredging projects is becoming increasingly difficult, farmers have applied various ad-hoc strategies, including individual requests for a machine to dredge a certain section of canal. This measure can supply water for irrigation of the farmer's field but does not help with transportation along the canal (interview 29.12.2011). This strategy creates a temporary solution for households with financial means but ultimately contradictory impacts. First is unequal water appropriation among farmers along the canal as water flows more to the deeper, dredged section and little or no flows the canal sections which are not dredged. However, rather than creating conflict, this inequality in water appropriation is justified by the capitalistic understanding that those with money to invest in dredging have the right to use more water.

While keeping an eye on such activity, the states realises the unsolved problems which create risks for agricultural production in both irrigation and trading (transportation through the canal reduces costs and increases traders' opportunities to reach farmers) and has implemented alternatives. Elsewhere, many field canals have dried up or can no longer connect, while other canals are growing larger (more than 10 m wide), leading to higher dredging costs. In this situation, the state, specifically the district state, does the following:

> At the moment, there is no distinction between the first, second and field canals anymore. In general, when there are canals that need to be dredged urgently, they [farmers and commune state cadres] will give suggestions. If there is a budget, either from district or people, we [district state] will bring the machine without deciding whether it is field canal, whether it is the responsibility of the people or of the [State] commune. The commune has no budget, while the people [farmer's], some have 2–3 cong (about 0.3 ha), their income is just enough to eat; there is no money left for dredging. (District's agriculture cadre, interview 19.04.2011)

The state agencies at district level often '*start with people mobilisation; if not, the district will try to invest*' (interview 14.12.2011). In other cases, district state officials first ask farmers to contribute labour or money for dredging. If farmers do not agree to dredge, the district state will try to find money for it. With recent, possible alternatives, the interviewed farmers and commune/hamlet cadres in the researched area tend to wait for state support, especially when the state shows signs of assuming responsibility for the field canal dredging budget. The possibility of external support and increasing costs has created expectations that have made some individuals hesitate to contribute to these projects. Dependence on state support has also arisen due to increasing local use of technology, in this case, dredging machines. As well, free-riders in canal dredging projects might wholeheartedly believe that it is the state's responsibility to fill the financial gaps in dredging. Above all, due to marketisation, the state has intervened by treating dredging as a monetary gift when it might have used coercion to organise dredging in the pre-Doi Moi collectivisation period. The contemporary state's capacity to use coercive powers thus is undermined by its activities as a helping state. In contemporary Vietnam, the state has shifted from authoritarianism to *corporatism*, uses mobilisation instead of coercion and allows *dialogue* and negotiation at the practice level in order to demonstrate state promotion of grassroots democracy. In this context, the local state has no power to force inhabitants to take actions and, thus, has failed to abolish free-rider problems.

In the contemporary period, the increasing challenges in local co-operation and contributions to canal dredging have forced farmers to search for ad-hoc solutions and have created space for a state presence in the community. However, the state does not desire this financial dependency. Again, with a limited budget, the state faces a dilemma of letting farmers handle practice level while ensuring the farmers are not left alone to cope with problems resulting in production losses, which can lead to the failure of the state's production and poverty-alleviation plans. The state intervenes only when the situation threatens food security or when the need to dredge a specific canal is urgent. Including field canals in the state budget for dredging, which is already short, means that the requirement to dredge canals every 3–5 years can rarely be met.

Conclusion: Negotiating co-operation for less urgent activities

Today's dependence on mechanisation in canal dredging has resulted from the general growing importance of machines and private businesses in agriculture and construction. Mechanised dredging can fill the labour gap left by outmigration from rural areas, has created a stable, new business sector for many actors and supports the ideology of technocratic solutions for development (Benedikter 2014b, Blake 2012). However, it faces difficulties in getting buy-in from resource-poor farmers.

Field canal dredging is different than field drainage and is an individualistic, capitalistic matter, making the conditions for co-operation even more vulnerable. Farmers enter negotiations over canal projects with their own views of reality (Rosen 1984) and decision-making power based on their financial capacity. The lack of economic interdependence between farmers results in few incentives for communal co-operation, which is based on cost–benefit calculations. In field canal dredging in the open Delta society where farming was returned to households in the 1980s, individuals have never fallen into the prisoners' dilemma (Ostrom 1990) and have more opportunities for free-riding actions as social constraints and state authority have little influence on their behaviour.

When the deterioration of the canal system might threaten food security, the state enters the sector and can undermine farmers' sense of self-management through its monetary support. However, state support is not a sustainable solution as the state consistently lacks sufficient finances, leaving many farmers to wait on the state budget for dredging. The case of canal dredging demonstrates the interdependence between farmers and the state. The state has succeeded in discursively legitimating its presence throughout society, giving it control over territory. However, by entering farmers' management sphere — field canal dredging — the state has failed to achieve its policy to retreat from financial and management responsibilities. Consequently, the inconsistent, short state budget also has to carry the burden of field canal drainage. Until another solution is found, the state will operate on a '*đến đâu hay đến đó*' basis (will fund when possible) (interview 18.10.2011). The state's funding of field canal dredging on a project by project basis is an experiment while it negotiates social arrangements for the sector.

Conclusions: local irrigation management as an institutional process and the result of state–society interactions

Focusing on the Vietnamese state's decentralised management structure, this chapter analyses the bureaucratic–informal interface of state policy implementation and local arrangements for field drainage. Field drainage activity provides empirical evidences of a complex process of practice that helps answer the question of how and why practice develops in the Mekong Delta. Briefly, the local practice of field drainage is guided by a complex process of interactions and negotiations among various stakeholders that combines the institutional formation of bricolage and an adaptive relation between the state and local actors.

Analysing the interaction between the state and local stakeholders in implementing state policy in field drainage and field canal dredging activities sheds light on how the state envisages its role in achieving control and management over production in the Delta, including field drainage. Based on evidences form the Delta, we propose the following conclusions about the Vietnamese state regime.

First, by using a standardised, hierarchal structure with agencies at 4 administrative levels to manage the country, the Vietnamese state faces the impacts of rigidity and redundancy. Standardisation applied to make society legible, on one hand, helps the state transfer policies and shape the current social arrangements at the local level where the state's presence and supports are needed. On the other hand, standardisation results in redundancy. For example, the meetings to disseminate the production plan from the province to the commune only transfers mandates, instead of fostering discussion of how to implement the plan. Additionally, the rigidity of the state structure prevents discussion of the suitability of standardised policies, which also tend to be too rigid for diverse localities.

Second, empirical evidences from field drainage indicate that, in applying its various powers, the state follows a soft–authoritarian approach, not a command-and-control regime. State policy implementation relies on the application of infrastructure power, with a hierarchical procedure for disseminating information and mandates through a dense network of state agencies that draws upon the power of modernism based on scientific knowledge and technologies. The current state management style differs from the pre-Doi Moi command-and-control system. Today, the Vietnamese applies soft control to impose rigid, bureaucratic mandates while allowing flexibility at the local level. Through the encouragement nature of agriculture policies, the state promotes grassroots democracy, Vietnamese style: freedom with mobilisation. Mobilisation efforts are part of the state structure of governing: applying discursive power and using political propaganda to shape the political and social realities and state correspondence to these realities. The current regime is based on the state's past experience that imposing a single model on a diverse society is not only unrealistic but also risks local resistance, both covert and overt. Therefore, policy-makers in the Delta prefer mobilisation strategies combined with allowing flexibility. These choices, according to some scholars, have resulted in a soft, dominating, corporatism state (see Kerkvliet 2001b and Kerkvliet 2003 on the types of Vietnamese states).

Third, the Vietnamese state is becoming more responsive in policy making and implementation, for example, in the incorporation of local farming traditions into the sowing schedule and the acceptance of deviation in implementation of the production plan. The rigid state mandate given with a bureaucratic checklist, in fact, turns out to be close to reality until the hamlet pumping meeting, when the real negotiation and discussion of field drainage practices starts. Although the state is becoming more responsive and is careful to avoid imposing a command system, it still seeks to simplify society. It continues to administratively order nature and society through various strategies, including mobilisation, state supports, encouraging policies and promotion of grassroots democracy.

Even as the state responds to local conditions, local stakeholders respond to the state. Over the years, the influence of the state system has made farmers familiar with the state's presence and, to a certain extent, dependent on its guidance and support. The establishment and increase of the inter-dependency between the state, its policy, local stakeholders and their practices determine the nature of institutions guiding the local practice of field drainage.

Negotiating irrigation management as forming of institutions

In field drainage, a set of institutions consisting of various rules and norms guides practice. These rules and norms are in constant negotiation over the validation or the suitability of a specific activity at a given moment in time. Supplementing and contradicting each other, those norms constitute the present institutions and so decide practice. Thus, institutions are complex, the legitimation of their components and norms always under negotiation. In certain contexts, the various pieces of norms together form a bricolage (Cleaver 2012). In field drainage, the norms or materials for bricolage are individualism, high social spirit, economic dependency and social tolerance and can be selected to form new bricolage when necessary. Thus, institutions in the form of bricolage are neither old nor new but a combination of both. In the search for alternatives to ensure fair, effective field drainage organisation, new materials are added and validated through trial and error. For instance, a commune cadre explains why the alternative of contracted pumping is favoured:

> We all agree with this model of contracted pumping in order to collectively avoid plant hoppers and reduce fuel costs. Hiring a big pump with a contract is cheaper than the individual pumps, and we don't need to take care of the pump all the time. The women especially like it [laughing] because the husbands, brothers and sons do not need to stay in the field to check on the pump day and night. Often, gathering in the field means drinking a lot, like in Tet [the annual lunar New Year which is the occasion that people gather and drink all days]. Sometimes, the wine costs as much as the fuel spent. (Commune cadre, after pumping meeting 17.10.2011)

Current practices and local institutions are maintained as the reasons for their existence remain, and the alternatives under negotiation have not yet formed another bricolage (Cleaver 2012). Zoning and making pumping groups support the state agenda of promoting a collective economy which overlaps with farmers' incentive to work together in collective pumping. The economic interdependence and benefits of collective field drainage, which are the most important factors in individual decision making, are still higher than the costs. Thus, co-operation in collective field drainage remains common, despite the persistent problem of free-riding actions. The strong influence of economic factors, though, has made co-operation more difficult in field canal dredging, which is an individualistic and capitalistic activity. The organisation of field drainage activities illustrates that the nature of the institutions in the open society of the Mekong Delta both constrains human behaviour and is negotiated and decided by humans. In negotiating arrangements, local cadres, zone leaders and farmers decide the choices for practice in certain contexts.

CHAPTER 6

NEGOTIATING LOCAL IRRIGATION MANAGEMENT IN THE MEKONG DELTA: STATE-SOCIETY NEGOTIATIONS AND BARGAINING IN THE FORMATION OF INSTITUTIONAL BRICOLAGE

The institutional setting and the state role in contemporary Vietnam have been among the exploratory interests of scientific scholars. While researches have already reviewed the state management structure as well as many issues regarding policy making and implementation, an empirical account of the local process of deciding practises is still missing. In filling the gap, this research aims to understand the local processes in irrigation management that involve not only the state agencies, but also other actors at the local level. This is primarily accomplished through anthropological research at the practical levels of field drainage and canal dredging. This research sets out to explain the local practises of field drainage and canal dredging in the Mekong Delta using the concept of institutions. The research also gives empirical evidence for the forming of institutions through the processes of interaction and negotiation at the interface, which is shaped by the bargaining of individual behaviours. Also, through analysing how local processes of interaction and negotiation guide the practise or the implementation of state policies, this research sheds light on the state's status and its use of state-related power in relation to the local actors, including the local cadres. This study seeks to answer these two questions:

- What are the informally agreed-upon elements of the platform of mutual learning and co-evolution between the state and local actors?
- How do practises of irrigation management respond to the 'everyday dialogue' between farmers, local state cadres and other state managers?

This chapter brings together answers to these research questions but also highlights various factors that expand academic knowledge of various concepts and of the region. It opens by illuminating the nature of institutional processes surrounding field drainage and canal dredging activities, whereby empirical evidence illuminates the decision-making process leading to co-operation, negotiation or the establishment of new arrangements for validating and legitimating institutions. In previous sections, this process was referred to as a kind of institutional bricolage, wherein decision-making is shaped by individual behaviours that are, in turn, informed by social, moral, emotional and economic factors. This book analyses local irrigation management at the interface of "bureaucratic-informal" in field drainage and canal dredging activities. Interactions at the "bureaucratic-informal" interface are designed by state bureaucratic structures and local arrangements guided by local stakeholders, including the local state cadres who are part of the local system. This interface, according to Norman Long and his colleagues in Wageningen UR, is the site for common interests, conflicts, incompatibility, and negotiation, in which "*Continued interaction encourages the development of boundaries and shared expectations that shape the interactions of the participants so that over time the interface itself becomes an organised entity of interlocking relationships and intentionalities*" (Long 2001:1-2). This research goes beyond the local negotiation of institutions and the bargaining of individual behaviours to look at the co-evolution between the state and local actors, including the state cadres. The last section of this chapter provides a study of state-society interactions through the application of power in various dimensions. Here, I draw attention to the nature of dialogue with the Vietnamese state – apparent control but provisions for flexibility – and within the co-evolution or the 'everyday dialogue' between the state and local actors in the market economy, privatised farming and interaction processes.

Irrigation practise in the process of negotiating institutional bricolage at the "bureaucratic-informal" interface

While the empirical evidence presented in this study more generally provides support for analytical pathways provided by critical institutionalism, the subsidiary concept of 'institutional bricolage' mostly seems to characterise the negotiation of local irrigation practises, specifically in field drainage and canal dredging. The utility of the institutional bricolage approach is not only observed when one evaluates the hydraulic interventions and the utility of farming technologies over the history of the Delta's water landscape, especially since the South migration in Nguyen's Dynasty (see chapter 3), but it also helps methodologically when combined with anthropological fieldwork methods in evaluating the dynamics of various day-to-day events (chapter 5).

As established in chapter 3, the contemporary water landscape of the Mekong Delta is the result of the *work in progress* of hydraulic interventions. Throughout the history of human migration, colonisation and nation-building since the 19th century, new infrastructures have been built on the foundations of pre-existing ones, and many investments and innovations in the past remain either totally or partly in the future landscape. Although each period of intervention did play a role, the post-colonial period with strongly technocratic-focussed projects shaped most of the contemporary features of the Delta landscape.

In the *work-in-progress*, the rulers interact with other actors (e.g., the armed opposition, the big farmers, the conscripted labourers, and others) under the influence of natural, physical, social and political conditions from which the results of periodic interventions were shaped. Since 2000 years ago, from the first recorded civilisation of the Funan people, to the periods of influence of Chenla and later the Kingdom of Khmer, to Vietnamese settlements, the colonial era, and the contemporary period, attempts to conquer nature have involved both successes and failures. In addition, the infrastructure and culture of the land over time have been under the social selection of policies and technologies to decide what to keep and what to demolish. The political factor has been obviously reflected in the persistence of military purposes in all projects, given the armed struggles between opposing forces throughout the country's history. While nature has either encouraged or constrained human interventions, the pre-existing infrastructure has served as the foundation of the new projects, the social and cultural rules and norms have negotiated and legitimated the periodic infrastructures and technologies, and the struggle for political rule has shaped the politicised interventions with separated zones and a mixture of construction and destruction.

The way in which the present Delta's landscape of infrastructures, technologies and cultures was formed under the influence of natural, physical, social and political factors reflects the institutional process in the history of forming the Delta's landscape. In this process, rules and norms have been negotiated as technologies have been legitimated. This has been illustrated by the reconstruction of the Cham's exploitative systems, the Khmer peasants, Chinese merchant-lenders and transporters such as crop cycles, irrigation subsystems, boating, fishing, and economic organisations. Also, it was the adoption and survival of mechanised canal construction that used machines instead of military or conscripted labour since their introduction by the French, and the Americans' mechanised technologies for agricultural production, specifically the utility of the water pump. Adding to the institutional process, the living and working culture of the Southern Delta is a reproduction of various cultures over the complex history of interactions with nature and between human settlements. As a consequence, it partly contributes to the diverse and open society of the Mekong Delta. The way that people negotiate and legitimate the rules and norms of certain arrangements is the process of the legitimation of institutions, or the process to qualify the new pieces to be contributed to the bricolage.

The bricolages described in the making of the Delta´s water landscape include the choices for technologies, hydraulic interventions and farming practises of the Mekong Delta throughout history. The Delta's contemporary water landscape includes age-old institutions such as a water-dependent lifestyle as well as more modern arrangements such as a hydraulic water landscape that facilitates agriculture and the most recent developments in mechanised water management using dredges and pumps. As such, the institutional process and historical approach help explain the dynamic and reason for the contemporary water landscape of the Delta, but how this also contributes to the understanding of the dynamic of the practises in the contemporary period. The following section, based on the discussions in chapter 5, will shed light on the issue by looking at the day-to-day institutional process and individual behaviours.

Local irrigation institutions: a cocktail of "modern" and "traditional", of "formal" and "informal"

This research proves that the day-to-day management of field drainage and canal dredging involves a process of interaction between the state and local actors in which processes, choices and arrangements are negotiated and choices or practises are decided (Figure 6-15). The choices are made by individuals who decide their behaviours in certain activities by considering economic costs and benefits, the moral norms that they have been familiar with, their own emotions towards certain arrangements based on past experiences, and their choices in relation to other individuals, including their neighbours and/or kin. The selection or negotiation of institutions from among various choices includes both the old one and the new one; the state-designed regulations along with the local common arrangements and the individual feasible options; both the traditional cooperative way based on "*village love and neighbour relations*" and the more modern market-oriented way of co-operation with accurate monetary calculation. When various pieces of rules and norms are chosen and applied in the practise of an activity at a specific time, the collection of pieces can be called an institutional bricolage.

In the contemporary Delta, at the local level, the bricolage of field drainage practises of the Thoi Lai district consists of pre-existing arrangements such as collective pumping, co-operation between farmers and the state based on goodwill or the high social spirits of individuals, as well as more strictly defined arrangements emerging from cost-benefit calculations, counteracting free-riding and slippage, and the availability of other alternatives. Thus, studying this bricolage can help provide an explanation of the dynamics of the Delta's water landscape. The following sections will substantiate this more comprehensively and illuminate the characteristics of the various institutional bricolages based on the empirical accounts from the Mekong Delta that have been reviewed during the course of this work.

Figure 6-15: Institutional bricolage at the interface of field drainage

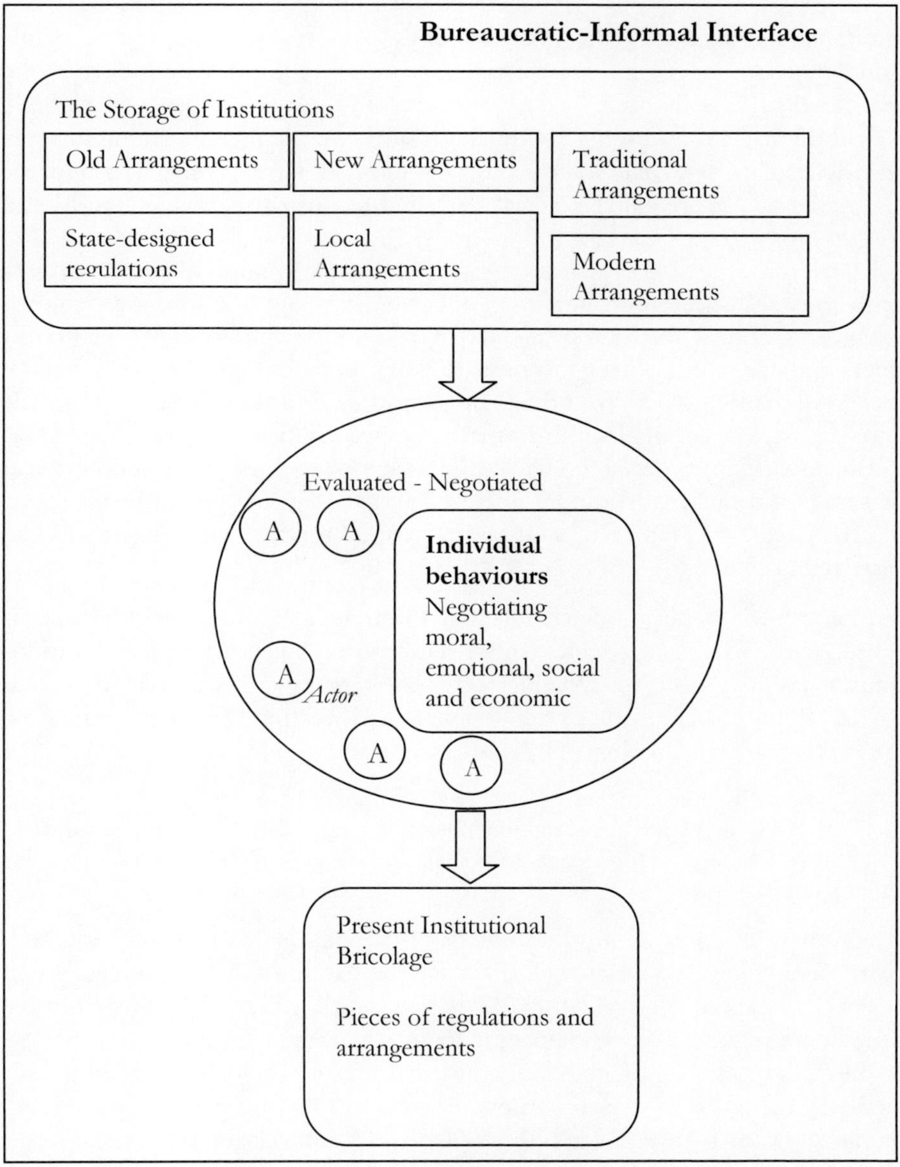

Based on Cleaver (2012) and empirical data from Thoi Lai district, Can Tho city.

In the context of local irrigation management, given that institutional bricolage aligns pieces from various arrangements, institutions are neither modern nor traditional, new nor old. The presently observable institutions, while maintaining some of the pre-existing and persistent rules and norms, are nevertheless under the influence of new rules and norms that are being tested. In field drainage, they are the combination of the newly formed and under-trial individualism and highly monetary dependency in farming activities and the old/traditional or pre-existing social rules and norms of social high spirits and working to increase harmony among neighbours. What are accepted and translated into practises of irrigation management today are both state-designed regulations and local arrangements. For instance, zoning or forming pumping groups is amongst the state's agenda of promoting the collective economy, which overlaps with farmers' incentive to work together on collective pumping. The combination of both formal and informal institutions as well as the boundary between formal and informal are especially blurred in this context, as explained within this research. The combination of multiple rules and norms at the same time causes institutions for local irrigation management in the Delta to become a "cocktail" of "old" and "new", "formal" and "informal" (as illustrated in Figure 6-15).

The nature of bricolage, which includes different institutions at one time, is typical of the Delta, in which human settlements have encompassed various cultures and are more readily able to accept new practises and adapt to old practises. This characteristic of the people in the Mekong Delta has been noted by various scholars. Brocheux (1995:91) asserts that:

> "Mien Tay [the Southern West area or the Mekong Delta] was a frontier; however, its people harbored individualistic tendencies. They were included to show less respect for recognized values and were more active in the quest for material goods."

Based on one explanation (Rambo 2005:15), this is because the Southern environment, with the advantages given by the nature of rich resources, posed far fewer threats to survival compared to the North, where high risks from the natural environment compelled people to arrange more corporate communities. Additionally, because of the individualistic lifestyle associated with people settled along waterways, Southerners created weaker social networks (Hy V. Luong 2003:18; Nguyen Ngoc De 2006:69-70) and villages remained loosely structured (Biggs et al. 2009; Jamieson 1995:5). In a note on the Delta's lifestyles, Son Nam highlights the Southern discipline that demonstrates the flexibility in which "*static culture is dead culture, active life adapting to conditions is culture in its full meaning*" (Son Nam 2003:56). In addition, because of this loose type of network, the people in the Delta are freer to organise than others

(Fforde and Nguyen Dinh Huan 2001:11). In fact, the Southern villages from the beginning did not stick to any form of organisation like the ones in the North. This, in fact, has posed difficulties to rulers who merely allowed the flexibility and freedom of organisation that were already the case. The Southern communes' freedom from regulation caused the disruption or poor implementation of collectivisation after the reunification in the 1970s. With the goal of searching for land to reclaim, the ones living in the South were always under pressure to keep an open mind and be quick to adapt and accept new conditions and cultures. In such an open society, the material for bricolage is richer.

The open society of the South indeed paves the way for new factors entering the institutional negotiation ground. As Hickey (1968:285) concludes, "*the villagers feel that the old ways are good, especially when they are expressed in rituals, but they also feel that the new ways are acceptable when proven of value*". As such, scientific innovations become localised by farmers if they are perceived as relevant to local problems (Marzano 2009:156). In this research, this was observed in rice cultivation practice, cultivating techniques, pest management, the acceptance of state sowing schedules and the recent trial of contracted pumping. Thus, besides the overall features of the institutional process that facilitate the formation of the institutional bricolage made from old and new and formal and informal institutions, the Delta's specific nature of openness has created a richer and more diverse institutional storage for bricolage and the environment for easier and faster adaptation or legitimation of various institutions.

Furthermore, current institutions continue changing through this process of negotiation and adaptation. Empirical accounts of the formation of the institutional bricolage in field drainage activity reveal that various arrangements are considered and negotiated by the stakeholders. Consequently, each arrangement is decided to be either dominant or to have only a minor appearance in a certain activity or a certain location. In this study, the negotiation of institutions in irrigation management happens in day-to-day encounters at public places such as a hamlet's information house, a farmer's house where a pumping meeting takes place, a market, a rice field or a cafeteria. In the process of choosing and adapting arrangements from a range of rules and norms, institutions are shaped and reshaped through the actions and interactions of the involved actors. This agrees with Cleaver (2012:182) that "*… it does not just happen; it is undertaken by people consciously and unconsciously navigating the institutional landscape and patching together "new" arrangements in a dynamic environment.*" Thus, an arrangement must be accepted and validated, and most of the time needs to be used repeatedly. The fight for legitimacy is continuous, meaning the accepted rules are under reassessment and might be replaced.

The past arrangements stay because their reasons for existing are still there. Some of today's arrangements that are age-old include the reliance of agriculture on hydraulic structures, the domination of dredges in the excavation and the maintenance of small and big canals, and the popular usage of pumps in farming and everyday life in the rural areas. In the irrigation context today, these and other similar practises are considered the best solution for farming at the moment while the new arrangements or alternatives are under trial and being adapted. Thus, the Vietnamese history of using or being affected by old arrangements confirms Cleaver (2012) statement: "*old arrangements have a remarkable tenacity, informing and shaping apparent innovations*". Speaking closer to the Vietnam case, Jamieson (1995:2) wrote that "*Vietnamese people are today still, as they were fifty years ago, interacting with that past in the process of shaping their future*". All in all, in this study, old arrangements that were valued and accepted continue to have a significant impact on the actors, both consciously and unconsciously, in the decision-making of all current practises. However, the valuation and acceptance of certain institutions does not necessarily derive from rationales about actual benefits; many times, it is decided by other factors such as pre-existing beliefs or the inertia of technologies and institutions.

The influence of beliefs about certain ideologies, e.g., a technocratic solution in building hydraulic works is the "best" solution (fast, easy and providing more benefits) for development, plays an important role in retaining old arrangements. As discussed by Benedikter (2014b) and Biggs (2011), the move towards mechanised and thus monetary dredging is becoming a beneficial sector for many actors and the ideology supporting it as a technocratic solution for development. Over the history of the Delta's landscape, it is the technological and institutional inertia that support the survival of certain technologies, management and intervention methods. In fact, analysing the past arrangement of relying on private contractors to carry out public works has been increasing since the 1880s, Biggs et al. (2009:221) assert that the large amount of interest through project lobbying and the domination of aging infrastructures allows no space for propositions to make major changes in water resource strategies. Similarly, the concentration of budget and management for canal dredging at province and district levels, as discussed in chapter 4 of this work, has created barriers for local correction and participation in the activity. The process of validating institutions that decide arrangements in practise thus involves the laziness of change, either because of the politicised institutions or the uncertainty potentially derived from the new arrangements.

The pre-existing arrangements thus continue to be practised, but are done so together with new institutions that form the broader institutional bricolage for irrigation management. This is how informal old ways of pumping together become an organised field drainage event formed and facilitated by the hamlet cadres or community leaders, in which, in some places, the organisation of field drainage has transformed into contracted pumping. The same tendency is recorded in the Red River Delta, where the water control has proven to be a mixture of old features from the collectivist period and new creations since the last two decades of liberalisation and decentralisation (Fontenelle 2000). Institutions in resource management are shaped both deliberately in formal spaces (including varying degrees of publicity and formality: committees, mass organisations, farmer organisations), and less consciously in routine daily interactions (embodied in kinships and social networks, relations of reciprocity and patronage, and in sets of norms and practises deeply embedded in the habits and routines of everyday life).

The application or legitimation of various arrangements is often endogenous, which is to say that the stakeholders match and adapt their mutual arrangements simultaneously. In this research on field drainage, the state adapts its production planning to be suitable to local practises of sowing at low water and from the beginning of October (lunar calendar), while local people make accommodations to the state's production planning, including the state sowing schedule. This confirms the dynamic adaptation in which people patch together their institutions from social, cultural and political resources (Smith et al. 2001:42, cited in Cleaver 1999:15). The extent to which various arrangements can be adapted to practise depends primarily on their utility (its suitability in time and space) and less on who initiates them: the local farmers, or the 'outsiders' such as the state or the scientists.

While local arrangements tend to mirror local needs more closely, technological and institutional inertia inhibit rapid adaptation. Local institutions that persist primarily due to inertia tend to be maladapted to contemporary conditions. For example, the fuel sharing and the calculation of payment for canal dredging that is based on *tam* (see chapter 5) used to be a monetarily equal cost among farmers given the social arrangements; however, given the increasing monetisation of society, this method no longer equates with the same sense of fairness. Therefore, this research suggests the overcoming of the 'myth' of local governance that local arrangements are automatically better-suited more than any universal arrangement. Viewing socially embedded institutions as necessarily 'better' than formal/organisational ones can often turn out to be simply a populist orientation that upholds and reproduces locally specific configurations of inequity and exclusion (Cleaver 1999:603). In addition, in a rapidly changing setting such as the Mekong Delta, the broadening of choices and the growth of more individualistic decision-making put all local arrangements under renewed scrutiny.

Although the process of legitimating institutions takes time, and technological and institutional inertia slow down the adaptation, institutional bricolage provides a solution for the problem of maladaptation. The gradual transformation of institutions, such as the adaptation of old arrangements into some new elements, or the retaining of some old norms together with problematic behaviours that have been induced by the present livelihood changes, helps temporarily to solve the current problems. In contemporary field drainage, the new norm of taking care of your own household in the market economy can lead to a lack of co-operation and free-riding behaviour. This action is tolerated and compensated for by the old or traditional high social spirit of doing something good for other individuals in a hamlet. The collective pumping in field drainage thus survives because of the combination of both ideologies when the fairer arrangement for pumping, like contracted field drainage, is not yet in place or legitimated. In the pool of materials for bricolage, individuals have to make choices for their institutions. The institutional bricolages, while creating diversity and limiting attempts to identify and manage, provide the most possible compensation for any negative impacts of both old and new regulations.

The local dynamic of the interaction and negotiation of policy implementation or the local practise of irrigation management have thus been explained by the institutional bricolage approach. In the cocktail of old and new, formal and informal, the bricolages increase the chances of adapting to new situations while a single set of regulations, either old or new, designed by the state or the local rules is not able to do so. While confirming the characteristics of institutions in the form of bricolage, such as the combination of various institutions and the continuous process of the legitimation of rules and norms, this research contributes two things to the discussion of the concepts. First, evidences from the field drainage and canal dredging add empirical account for the formation and legitimation of bricolage. Second, bricolage in local irrigation management highlights several features of the Delta's institutional process, including rich storage for bricolage due to the open society and quickly adaptive features of the Delta; in the assessment and legitimation of pieces for bricolage, the old institutions stay because of their value of utility or the technological and institutional inertia. This research highlights the problem-solving role of bricolage in providing temporary solutions for effective local arrangements when adaptation is slow or in facing the challenges of new institutions (e.g., the reduction of value of social goodwill in the free-riding issue). The legitimation and valuation of social rules and norms that decide the negotiations are dependent on individuals with multiple identities. The following section will argue that individual behaviour is the result of a bargaining process that is shaped by various internal and external drivers.

Bargaining individual behaviours in negotiating the institutional bricolage

The process of forming institutional bricolage includes interaction and negotiation at an interface where stakeholders adjust power relations. As institutions form, they guide the practises by both encouraging and constraining the behaviours of those involved. In negotiating institutions, each individual goes through a process of bargaining the reality which is to decide their behaviours in practise. Empirical accounts from day-to-day negotiations in field drainage and canal dredging illustrate that each individual with multiple identities possesses different levels of power in bargaining their behaviours and negotiating the arrangements; in other words, each individual enjoys various kinds of "room for manoeuvres". The following section will synthesize the analyses, mostly elaborated on in chapter 5, supporting the argument while proving the influence of various internal and external factors on the bargaining of self-behaviour.

The diversity in behaviour towards pumping co-operation in field drainage implies the imperfect influence of pre-existing rules and norms on individuals. As Cleaver (2012:119) states, "… *whilst individuals are embedded in social life, they are not totally submerged in it – structures, rules, norms and relationships are all mediated by human agency*". Rather, "*the effects produced are uneven due to the patterning of social structures and the dynamism and unpredictability of inter-connected human actions*" (McNay 2000, cited in Cleaver 2012:116). In the researched site, those in the same commune, hamlet or even in the same zone are all under the influence of established institutions such as following the state guidance for production planning with sowing schedules, acting on goodwill and maintaining good relations with one's neighbours. However, the extent of influence varies between individuals; each individual has the power to decide and may choose to behave differently in negotiating the pumping date, in bringing pumps to the field, in paying for the fuel, and in helping with dyke repair during pumping. As a result, while some zones manage to pump and sow according to state or scientifically recommended plans, the other zones either sow later or are divided into smaller zones if an agreement cannot be reached. Thus, the diversity at the local level or the unpredictability caused by human agency in re-negotiating institutions makes it difficult for the state to predict the outcome of a policy.

In such diversity of the local level, the possibility of failing to implement the general policy of the state might be partly compensated for by the local dynamic. The variations in responding to social rules and norms between individuals result from the nature of individual behaviours, which are informed by moral, emotional, social, political and economic factors. Cleaver (2012) summarises that, in general, the behaviour of each individual is driven by internal factors (cognition, emotion, ideology) and external ones (e.g., social relationships). In negotiating field drainage activity, farmers do care about social norms as it is morally proper to help your kin and your neighbours, and to respect the prestigious social (farming practise) leaders (ideology). In addition, the resentment from past experiences of following state guidance, co-operating with neighbours, or working as a group leader all affect one's behaviour (emotion). While one is going through the mental process to think, remember and judge what is allowed and whether they like certain rules and norms – under the influence of internal factors, individual behaviour is also affected by social requirements such as keeping up relationships with the neighbours or maintaining a certain 'face' in public. Such requirements or social rules are felt throughout the interactions in meetings or at markets, cafeterias, and rice fields, as observed in this research. This research agrees with Cleaver (2012:15) on the power of behaviour in social relationships and internal emotions and ideologies, as stated in the strategic livelihood choices model in which human beings are "*... critically influenced by social concerns (such as the need to live in peace with neighbors), by psychological preferences (for example, for cooperation over confrontation) and culturally and historically shaped ideas about the right way of doing things.*" In addition, farmers' thoughts are affected and can also be changed by state policies and mobilisation strategies. So, in the context of irrigation management, I include 'policy' as another external factor that guides individual behaviour. This factor is especially important in the local context of this region, where the state's deconcentrated structure and the policies have grown to be acknowledged as part of the local context. Thus, human behaviour is not simply encouraged and/or inhibited by pre-existing institutions including the social rules, the state-structured regulations (external factors), and the social norms that are unconsciously acknowledged by individuals (internal factors); the internal process of cognition together with moral and emotional factors drive behaviours, and thus the decisions of each individual. While internal drivers stay active in oneself, the influence of many external factors including the state influence and social rules and norms become obvious during the interaction between actors; and, at many times, one is compelled to include or adjust the internal factors (e.g., emotions) to accommodate the external contradictory driver(s) of behaviour.

Besides previously mentioned drivers of individual behaviour, the current context of the Mekong Delta adds another: the economic factor. Economic or monetary calculation may encourage or prevent one from contributing to pumping events or fuel sharing while monetary inter-dependence in dealing with threats adds a strong incentive for individuals' participation in collective pumping and willingness to help with dyke repair. Similar cases of the influence of economic factors are described by Scott (1985) and Boelens (2008), and it is concluded that individual behaviour is informed by internal emotional, moral, or social rationalities, in addition to economic factors. In a market economy with increasing costs of farming and living expenses, economic factors are becoming increasingly important in many previously non-monetised transactions.

At the hamlet level, none of the stakeholders acts according to any specific pattern of behaviour for the class or role of farmers or local cadres; rather, a person's behaviour is strongly driven by his/her multiple identities. Giddens (1984) asserts that a person is positioned in multiple ways with social relations conferred by special social identities, and Long (1992) concludes that each individual tends to become involved in several projects of social relations with rules and norms (cited in Cleaver 1999:606). In this way, a farmer who is at the same time a cadre (as described in Chapters 4 and 5) is involved in at least two projects: the local state project managing the sector, and the farmer's project of taking care of his/her farming production. From time to time, local cadres face difficulties deciding between state mandates and local requirements. As an individual involved in the farming and social 'project' with other neighbours and the state, one is also embedded in the household 'project', in which he or she is responsible for the family's income. Depending on the relationship between the two arenas in terms of interests and targets, the choice is made by the individual in order to maximise the benefits and minimise the costs. In the case of irrigation and rice production, the two projects are likely to overlap, and, in the case of contradictions, the possibility for co-adaptation by both state policies and local arrangements compensates and helps the individual to solve the internal contradictions. Because of potential conflicts in identities that are not easily resolved, individuals are left with the strategic choice to either (a) adjust their own behaviour, or (b) attempt to influence the external expectations of them.

While being driven by various factors from social relations, state policies and the internal factors of emotion, morality, and ideology, individual behaviour involves constant negotiation or the bargaining of power to decide the legitimacy of arrangements, and good logic will be accepted, which is one of the important characteristics of 'prevailing' to change local behaviour. Popkin

(1979:22) asserts that norms with unresolved inconsistencies and conflicts cannot directly and simply determine actions, but are more likely to be assessed though decisions are not easily arrived at or maintained. In the Mekong Delta, while one can be just a farmer, another may be both a local cadre and a farmer or a private businessman/woman; one may have more capacity in finances or materials or farming tools than the others. This confirms Cleaver's statement (2012:xii) that each person is socially located (with overlapping identities and motivations) and physically embodied. The identities shape individual capacities and motivations to act within collective action. Also, each has a specific view about certain contexts and logic and suitable rules and norms. This indeed reflects Rosen conclusion (1984:42) that, of several actors who are different from the others based on ethnicity, occupation or family background, "*each of them processes a different view of what is really true about the situation in which all are involved and attempts to make his or her definition of the situation prevail*" – forming the notion of reality bargaining (ibid). The process of bargaining the value and legitimacy of the influencing internal and external factors is the main driver of changes and deviations in local behaviour. Each consciously bargains the social relationship regulations, the state policies and the cost-benefit calculations for each arrangement, and less consciously negotiates the reality based on the emotions and embedded norms perceived in one of their own multi-identities.

Under the influence or the force to consider, consciously or unconsciously, various factors, individuals engage in a bargaining process which is based on a broad cost and benefit calculation. What is happening in the field drainage activity is similar to what described by (Ostrom 1990:30):

> An individual´s choice of behaviour in any particular situation will depend on how the individual learns about, views, and weighs the benefits and costs of actions and perceived linkage to outcomes that also involve a mixture of benefits and costs.

Indeed, costs and benefits from a certain activity considered by an individual are not only money-based, but are also social and (everyday) political costs and benefits. In the researched context of irrigation management in the Delta, the benefits gained must include social harmony. This confirms the previous discussion by Bauer (2011:166-167) that "*all rules and standards act under the principle of maintaining social harmony*", which is associated with peace, unity, social satisfaction and material prosperity, and thus encourages the tendency of Vietnamese society to avoid losing face. Also, one's own face or respect from farming neighbours is valued as important in everyday politics at the local level. However, due to current increases in the cost of farming and living, a situation that is typical for such a society is asserted by Uphoff et al. (1990:26): "*costs are usually definite and certain, whereas the benefits from participation in group action are often ambiguous and even uncertain*"; the market economy requires farmers and cadres to re-calculate different factors. As a result, the needs of individuals for recognition, respect or purpose are never independent from the material benefits, thus motivating the whole game of cost-benefit calculation in today's rural irrigation and production.

With the growing market mentality or the growing importance of monetisation in cost-benefit calculation, the bargaining for individual behaviour may lead one to the choice of free-riding. Free-riding issues from field drainage and field-canal dredging are either done consciously to provide less and gain the same, or unconsciously cause higher costs to the others by not co-operating in the joint activities. Above all, among variations in land-holding, financial capacity, and emotion, one rarely values economic and social costs and benefits the same as the other. The economic orientation plus the loose relationship in the Delta's society of private farming has provided the foundation for the decline in the consistent valuation of social good.

Within the dynamic of institutions, the state regulation is part of, and thus under the same procedure of, legitimation in which individuals either accept or reject each part of the regulation. Learning from the past, the state today with its embeddedness at the local level actually aligns institutions with their view of regulation. The state indeed acknowledges the local practise and local deviation policy implementation that is suitable to local conditions. By acknowledging the local flexibility, the state to a certain extent accepts the local process of legitimating the regulations or aligning their regulations with the idea of institutions. However, this process is not official and the state does not simply leave it; rather, with mobilisation strategies in the negotiation process, the state keeps trying to control and influence to make sure their regulations are accepted.

My findings suggest that policy makers should keep a few key principles in mind when aiming for appropriate regulations:

- State regulations must be formed from local requirements and scientific evidence. This is to maximise the suitability of state policies and local institutions, thus increasing the chances of policy implementation to meet state macro-targets.
- Given the local diversity, state regulations should officially allow flexibility, i.e., in identifying zone boundaries, in pumping and sowing dates, and in whether collective pumping is necessary. This research implies that allowing flexibility does not necessarily pose the risk of undermining state power; rather, with the local legitimation of state regulations in which process the state modernism power gained through scientific knowledge

and the growing adaptation of state policies to local conditions ensure the possibility for state policies to be considered and accepted, local flexibility compensates any shortcomings of the state policy.

All in all, individuals are constantly negotiating institutions through bargaining their own behaviours, which are under the influence of internal and external drivers. When interacting with the social context and with other stakeholders, individuals bargain for their power to decide, their status in the community and the legitimacy of rules and norms. In market-oriented farming life, economic factors are becoming dominant in the process of bargaining the reality in which material and social values are connected in cost-benefit calculations. Subsequently, in the given market orientation of relations, local cadres and farmers are facing the reality of monetisation and the growing importance of economic factors in all farming and irrigation activities as evidenced by the rise of the free-riding issue. Thus, in attempting to maintain some of the old norms of social good and social high spirits, the state continues implementing policies and mobilisation strategies to achieve influence and is gaining more status as a helper of farmers. The current inter-dependence between the state and local actors is the result of the process of mutual learning or co-evolving in refining the local governance. That process, which will be discussed in the following section, has formed the 'everyday dialogue' nature of the state-society interactions in the context of irrigation management in the Mekong Delta.

The state-society relationship in the negotiation of institutions at the "bureaucratic-informal" interface

In the presence of various actors, of whom the state is one, the local irrigation management in the Mekong Delta is captured by a complex structure of official deconcentration in which macro-division embedded in the legal framework provides guidance for the enforcement and implementation of the policies; and the local unofficial decentralisation, which is the result of ongoing interactions at the 'bureaucratic-informal' interface, shapes the local activities. The state and its role and status is captured in this research, not only as a primary actor, but in its relations with the other actors. Through the process of negotiation and institutions that are shaped by the bargaining of individual behaviour, this research reveals the local irrigation management though the 'everyday dialogue' between the state and local actors, including the local cadres.

In the contemporary Mekong Delta, the state has managed and organised the irrigation and agricultural production sector with the adaptive attitude of being more responsive to local institutions, including other rules and norms, but still desperate to hold control. Thus, the state's governance style or mechanism has included the application of state power in all three dimensions described by Göbel (2011): despotic, infrastructure and discursive power. Through the process of embeddedness in applying state power, the state becomes involved in a dialogue at the "bureaucratic-informal" interface where the state agencies and other stakeholders negotiate the policies together with other local arrangements, and, subsequently, practises are decided. This research suggests that the state and the local people (including half-state cadres) are in the process of learning from each other, such that they are both getting more effective at tailoring policies and their implementation, and handling their farming production.

The current Vietnamese state is usually depicted as a hierarchy of agencies from the central to the provincial, district and commune levels. Decentralisation in Vietnam in general stops at deconcentration that is based on task division and top-down mandates while devolution has been limited to the provincial level; there is an accumulation at the province and district levels of power over the administration and budget. The partial deconcentration resulting from this arrangement provides a persisting dilemma for the central government: how to give enough power to the local level for the state to retreat from financial and administrative burdens, while not undermining the state's role in the management and control of local affairs. This problem emerges distinctly in the issue of benefit over monopoly power in big infrastructure projects, such as budget allocation for canal dredging (described in more detail in chapter 4). Historical and empirical analyses demonstrate that the state structure and management have been based on the use of despotic, infrastructure and discursive dimensions of power (following Göbel 2011 identification of the power dimensions in an authoritarian state). In the context of irrigation management in the post-Doi Moi period, the Vietnamese state is mostly based on infrastructure and discursive power to engage in negotiations with other stakeholders; during the process of embeddedness, local stakeholders express their power of accepting and/or adjusting the policies (Figure 6-16).

Figure 6-16: The state-society relationship in the negotiation of institutions at the "bureaucratic-informal" interface of irrigation management

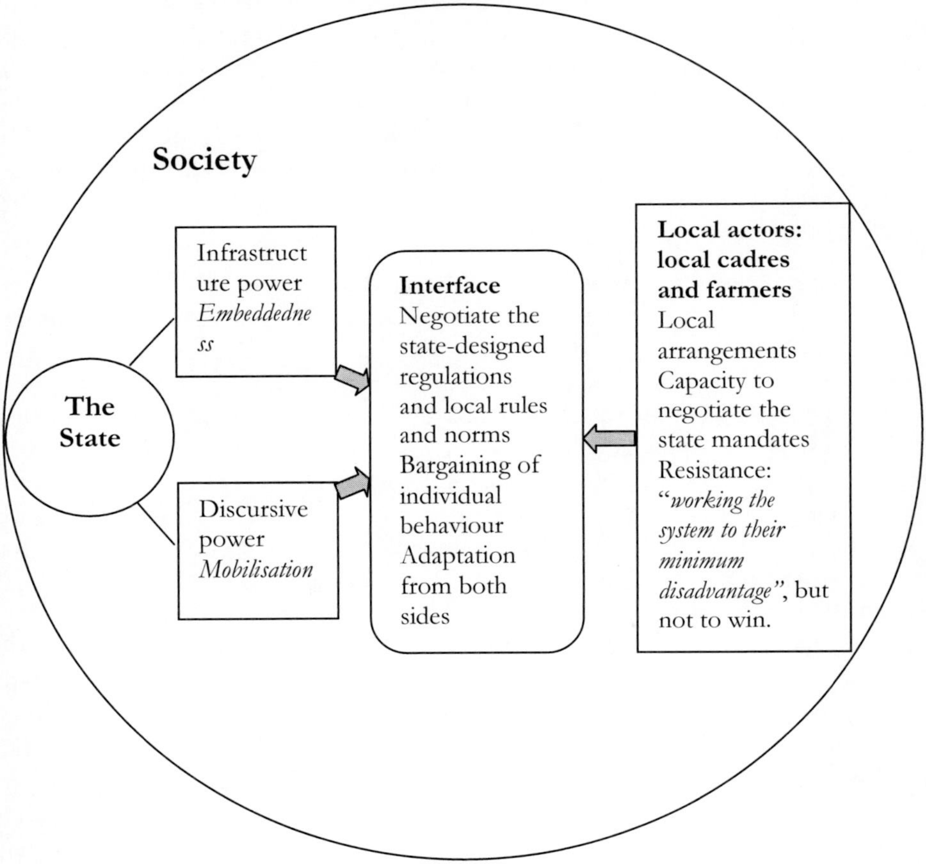

Since Doi Moi, the Vietnamese state, either by force or internal compulsion, has grown to be more responsive. The state was forced to abolish the collectivisation policy in order to make provisions for local practises of farming: currently, in the sowings schedule, by tolerating flexibility in arranging the local practises that are suitable and feasible for the natural, physical and social conditions of the specific localities. Today, in adopting a market economy, the state, while trying to attract more private and individual contributions in all activities, strategically retreats from those spheres. Allowing other agents to get involved invariably distributes influence and places the state in a defensive position. While state monopoly power remains in certain aspects such as politics and financial management, the state joins in negotiations with local cadres, farmers and businessmen/women in which practises are subsequently decided. In fact, due to the nature of the contemporary institutional bricolage, the state is decreasingly involved in people's lives in a direct or authoritative manner. Rather, state policies are increasingly assessed according to their suitability for the ecological conditions and social institutions. They are inserted into a more competitive process of legitimation with the participation of the policy-level state (provincial level in cases of field drainage and canal dredging), local state and affiliated agencies (commune People's Committee, hamlet People's Board, mass organisations, and many farmer organisations), private actors (pump owners, dredge owners), and farmers, as discussed in chapter 5 for the case of field drainage. This confirms the statement of Cleaver (2012:173) that the state engages with locally generated meanings and configurations of authority at the arenas of institutions where the state's formal rules become material for the practises of bricolage and the state agencies become bricoleurs joining other bricoleurs in negotiating and forming the institutions. Therefore, the state is now rarely the primary driver of the management, being merely a strong party in bargaining for favourable institutional conditions.

The state and other actors, including the local state cadres, have engaged in an inter-dependent relationship. This inter-dependence has been observed in the way that the state relies on local actors to achieve production targets while the local cadres and farmers rely on state support to run their agencies and enhance farming production. What is happening in the Delta's irrigation is similar to Göbel (2011:187) statement that the rulers and the citizens, both of whom are part of society, are connected through a relationship of belonging and believing. Such a relationship is shaped by history, the embeddedness of bureaucratic institutions in society, symbolic and material factors, and claims to legitimacy that decide the results of negotiations of statehood (Hagmann and Perclard 2010, cited in Cleaver 2012:172). This implies a process of mutual learning among both the state and other actors over time which is shaped by agreement and contradiction, adjustment and adaptation.

In becoming part of the local negotiation of institutions, the state has no better choice than to apply discursive power through mobilisation strategies. Empirical accounts from field drainage and canal dredging illustrate that the state regulations are not dominating the practise, but rather being put into a competitive game with other arrangements or institutions from which they are negotiated and legitimated. For instance, the state's sowing schedule and the state's mode of farmer organisation, as analysed in chapter 5, are some of the state-designed regulations that have been negotiated at the bureaucratic-informal interface. As the results of institutions, state-designed arrangements only become embedded when their value is proven. Through the process of embeddedness, the state also allows its adaptation and flexibility to other ideologies, rules and norms at the local level. The adjustment of policies is not about the outcome only, but about how the adjustments are represented discursively. If the state allows farmers to delay sowing, one can interpret the state as 'weak' or the state as 'wise and accommodating'. Thus, it is all about how one wants to be represented.

In any cases, the dense structure of agencies with state cadres and state associates maintains a state presence in many local affairs. Yet, even while attempting to be omnipresent, the state does not necessarily wish to control or be responsible for everything. Indeed, the state seems to be happy to withdraw from responsibility for local-level infrastructures, especially field canals, and other micro-arrangements (e.g., for collective pumping). Based on my work, I note that the state tends to define a threshold at which it no longer tolerates local deviation; below that, the state wishes to mobilise the local contribution of resources and responsibility in compensating to what is left by state retreat. The state has been trying to find a balance between handling responsibility but not losing control, reducing management burdens but maintaining the state's status of being there for help. In the new era of private farming with the requirement to be flexible and responsive and the necessity to reduce the state burdens in finance and management, the state's application of infrastructure power has been adapting to achieve the new targets of sufficient management because of embedded cost resources and too little cost legitimacy.

This book is about the day-to-day hamlet level and district level organisation, which the state has retreated from. At that level, the local actors have responded in various ways to the state regulations in the process of negotiating the practise. This research goes beyond the idea that the South of Vietnam has ecological advantages and loose community cohesion compared to the North, which gives the people little incentive for open confrontation with state apparatuses (Hy V. Luong 2005:141-142). Rather, what has been reported in the Delta since the reunification is not rebellion, but they are not silent 'fence-breaking' either. The case of the survival of few cooperatives before Doi Moi in the South was an open action but not a confrontation with the state policy. In field drainage, as analysed in this research, farmers, mainly with the consent of a commune and/or hamlet's cadres, unofficially make a selection of what policy to follow, and, if necessary, they attempt to amend or ignore it and continue their comfortable practise. The form of resistance happening in the Southern Delta is often open but soft resistance in single events and activities from separated localities, which can be called "*working the system to their minimum disadvantage*" (termed by Hobsbawn, cited in Scott 1987). Thus, from the inside of negotiation space, the local actors do have the chance and capacity to impose power leverage; however, with no interest in being against the state or for the growth to be dependent on the state, those local cadres and farmers choose not to win but to maximise their advantage from the negotiation.

Even with the capacity to do so, the local cadres and farmers today have little reason to impose any confrontation. Besides the adjustment in policy implementation, and sometimes the decision to contradict the state mandate, local actors, including commune and hamlet cadres and farmers, do not want to go against the state and its management in general. In a research in Malaysia, Scott (1987) asserts that the sub-ordinate classes are far less interested in changing the larger structures of the state and the law; rather, they use ordinary weapons such as "*foot dragging, dissimulation, desertion, false compliance, pilfering, feigning ignorance, slander, arson, sabotage and so on*" (ibid). In the contemporary Delta, the local people indeed have no interest in changing the state management structure, but this is far from similar to what Scott reported in that the sub-ordinate classes lack a realistic possibility of redressing their situation, so they have little choice but to adjust, as best as they can, to the circumstances they confront daily (Scott 1987:246). The Delta's farmers and local cadres are both encouraged and constrained by state mandates, the incentives to enhance production income, and the risks of farming failure or higher costs. As such, with the growing recognition of various factors that directly affect farmers' decisions in farming, the farmers feel the courage to create or change, for instance, the decision to form pumping groups by farmers themselves (as reported in the Dong Thuan commune). This research indeed illustrates that the unclear and/or sometimes strategic task division in the state management of groups and co-operatives has given room for manoeuvres. Within that pocket of manoeuvres, the farmers have less fear of finding and trying alternatives, and local cadres can join and/or legalise those alternatives. At the "bureaucratic-informal" interface, while various diversity or negotiation results are possible, the co-operation required for certain farming practises provides an incentive for sacrificing freedom to neighbours or the state. In this mixture of constraints, incentives, and risks, farmers are working through all opportunities and problems that are also included in the state policies to achieve the most benefits at a minimum cost rather than criticising the state's political status.

With individuals farming within a new mode of state governance, individuals are considering all factors, including state guidance and other local natural, physical and socio-economic conditions, while going through internal moral and emotional situations to decide what will be done in practises. Similar to how the state might be represented differently, farmers' consideration of state policies can also be represented as a soft response to state policies, or the capacity to make decisions from a pool of choices and influencing factors. This research reveals a situation where farmers adjust to state policies, and, vice versa, the state has adjusted its policies according to local requirements. So, within the 'system', the state and local actors are involved in creating 'endogeneity' within state policies and local requirements, or they co-evolve.

Rather than to win, both the state and other actors, including the local cadres, choose to apply dialogue strategies in day-to-day interactions and negotiations. In cases where state policies would be more effective if implemented without coercion, the state relies on its mobilisation bodies (local, state, extension offices, and mass organisations) to facilitate the penetration of state policies while promoting grassroots democracy (people know, people do, people discuss, people assess) and allowing room to manoeuvre and freedom to choose. While applying the discursive power in order to make the people want what the state wants them to want, the state is also influenced by local dynamics which make the state evolve, or, in other words, the local discursive power. At the interface, the state joins a dialogue with local actors as the policies become part of, not the dominating factor in, deciding the practises. In terms of governance, what the state in Vietnam is practising is a form of 'meta-governance'. According to Bell and Hindmoor (2009), meta-governance not only expresses the capacity of a state in strategically adapting to a new context, but also highlights the dialogue-driven nature of a regime. With the new mode of governance in a new situation, the state has to amend its command-and-control system to a system where the state is in contact with various agents, from businessmen to local inhabitants in economic management and development. Since Doi Moi, the Vietnamese state has been compelled to grow more responsive. In doing so, the state is not necessarily exhibiting weakness; in fact, it can be considered strong in the sense of having the capacity to apply other modes of governance strategically, and by switching more fluidly between the three power dimensions: despotic, infrastructure and discursive. By accepting private economy to a certain extent, the Vietnamese state recognises and allows for dialogue where covert and/or overt but unofficial negotiations take place.

All in all, although the Vietnamese hierarchical structure imitates the so-called authoritarian regime, the dense structure of agencies from the central down to the hamlet that works with a top-down system of policy and mandates does not necessarily mean the system is not flexible. Rather, with empirical evidences from the cases of field drainage and canal dredging, this research exemplifies the whole relationship or interaction and negotiation between the government and other actors in the Mekong Delta as a complex situation in which there is a co-existence of hierarchical state management structure and the space of local flexibility. Today, the state, in finding and maintaining the balance between the state's retreat and its urgency to be present and state legitimacy, decides the threshold more than which it no longer tolerates local deviation. In other words, the local cadres and farmers, with multi-identities and engagements in various 'projects' at the same time, are able to accept, adapt, and/or adjust to the policies to be suitable to local conditions. Such room for manoeuvres is allowed by the state as long as it does not undermine the state power. Thus, in the current system, local flexibility allows local innovations and ideas to be tried with the possibility of being recognised and legitimated by the state. Such a top-down working style with flexibility is neither a democratic nor an authoritarian style that one can simply assign to the Vietnamese state.

This research goes beyond the local negotiation of institutions and the bargaining of reality of individuals' deciding behaviours to argue that irrigation management at the local level is guided by the process of co-evolution or mutual learning between the state and local actors, including the local cadres, in process of interaction, which the local governance has refined officially and unofficially while deciding the current practises of local irrigation management. While exploring the classic state-society theme, I turn to approaching the state as a part of society. Going beyond the idea of state-in-society and the blurred boundary between formal-informal in Vietnam as recommended by various scholars, not limited to Kerkvliet (2001), Gainsborough (2010), Reis (2010), the Heinrich Böll Foundation (2005), and Hicks (2005), this research on the interaction between the state and local actors at the "bureaucratic-informal" interface validates the nature of everyday dialogue in which institutions are negotiated in the process of forming bricolage. The negotiation process is guided by the bargaining of individual behaviours.

Within the dialogue, the state is not the primary driver, but is a strong party in bargaining favourable institutional conditions. Under the present state, success in penetrating into the social processes, both consciously and unconsciously through the application of infrastructure and discursive power (legislation and mobilisation), the local irrigation case of the Mekong Delta demonstrates the nature of local responses or the so-called local resistances, which is far from either rebellion or silent 'fence-breaking'. Rather, the farmers and local cadres with the constraints, incentives and risks perceived from the bargaining and negotiation process are working through all opportunities from both state regulations and local arrangements. In the competitive negotiation to make their own regulations and targets accepted by the others, the state and other actors have evolved into a learning process, as illustrated in the overlapping of interests and agreement in certain flexibility and deviations between places. This learning process or the process of negotiation has shaped the institutional bricolage from the various rules and norms that are valued and accepted differently amongst the state, the local state cadres, the semi-state actors, and the farmers.

References

Anonymous *Sub Areas in the Mekong Delta.*

Bauer, T. (2011) *The Challenge of Knowledge Sharing: Practices of the Vietnamese Science Community in Ho Chi Minh City and the Mekong Delta.* ZEF Development Studies 17. Münster: LIT Verlag

Benedikter, S. (2014a) "Extending the Hydraulic Paradigm: Reunification, State Consolidation and Water Control in the Vietnamese Mekong Delta after 1975". In: *Southeast Asian Studies, http://englishkyoto-seas.org/* 3 (3).

Benedikter, S. (2014b) *The Vietnamese Hydrocracy and the Mekong Delta: Water Resources Development from State Socialism to Bureaucratic Capitalism* ZEF Development Studies 25. Viena: Lit Verlag.

Benedikter, S. and Waibel, G. (2013) *The formation of water user groups in a nexus of central directives and local administration in the Mekong Delta, Vietnam.* ZEF Working Paper *Nr. 112.* Bonn: University of Bonn, Center for Development Research. http://www.zef.de/fileadmin/webfiles/downloads/zef_wp/WP55_Waibel.pdf. (22.04.2010).

Beresford, M. (2006) "Vietnam: The Transition from Central Planning". In: G. Rodan; K. Hewison; R. Robinson (Eds.): *The Political Economy of South-East Asia.* Oxford / New York: Oxford University Press. 197-220.

Bergstedt, C. (2012) *The Lie of the Land: Gender, Farm Work, and Land in a Rural Vietnamese Village.* Doctoral book. University of Gothenburg, Faculty of Social Sciences.

Biggs, D.; Miller, F.; Hoanh, C.T.; Molle, F. (2009) "The Delta Machine: Water Management in the Vietnamese Mekong Delta in Historical and Contemporary Perspectives". In: F. Molle; T. Foran; M. Käkönen (Eds.): *Contested Waterscapes in the Mekong Region. Hydropower, Livelihoods and Governance.* London: Earthscan. 203-225.

Biggs, D.A. (2004) *Between the Rivers and the Tides: A Hydraulic History of the Mekong Delta 1820-1975.* Doctoral Book. University of Washington, Department of History.

Biggs, D.A. (2011) "Fixing the Delta: History and the Politics of Hydraulic Infrastructure Development and Conservation in the Mekong Delta". In: Mart A. Stewart; P.A. Coclanis (Eds.): *Environmental Change and Agricultural Sustainability in the Mekong Delta.* London, New York: Springer. 35-44.

Biggs, D.A. (2012) *Quagmire: Nation-building and nature in the Mekong Delta.* Washington: University of Washington Press. 320.

Blaikie, N. (2010) *Designing Social Research.* Cambridge: Polity Press. 298.

Blake, D.J.H. (2012) *Irrigationalism – the politics and ideology of irrigation development in the Nam Songkhram Basin, Northeast Thailand.* Doctoral Book. University of East Anglia, School of International Development. 416.

Boelens, R. (2008) *The rules of the game and the game of the rules. Normalization and resistance in Andean water control.* Doctoral Book. Wageningen University. 573.

Brocheux, P. (1995) *The Mekong Delta: Ecology, Economy, and Revolution, 1860-1960.* Madison WI: Center for Southeast Asian Studies University of Wisconsin-Madison. 269.

Brouwer, C.; Goffeau, A.; Heibloem, M. (1985) *Irrigation Water Management: Training Manual No. 1 - Introduction to Irrigation.* Rome: FAO - Food and Agriculture organisation of the United Nations. http://www.fao.org/docrep/r4082e/r4082e00.htm. (04.11.2014).

Bucx, T.; Marchand, M.; Makaske, B.; van de Guchte, C. (2010) *Comparative assessment of the vulnerability and resilience of 10 deltas.* Delta Alliance report *Nr. 1*. Delft-Wageningen: Delta Alliance International.

Bui Huy Dap and Nguyen Dien (1996) *Nông nghiệp Việt Nam. Từ cội nguồn đến đổi mới - [Agriculture in Vietnam. From the Past to the Present].* Hanoi: National Politics Publisher.

Burton, M.; Wester, P.; Scott, C. (2002) "Safeguarding the needs of locally managed irrigation for the water scarse Lerma-Chapala river basin, Mexico". Paper presented at *International Commission on Irrigation and Drainage, Eighteenth Congress,* Montreal. 15.

Campbell, I.C. (2012) "Biodiversity of the Mekong Delta". In: F. Renaud; C. Künzer (Eds.): *The Mekong Delta System - Interdisciplinary Analyses of a River Delta.* Heidelberg: Springer. 293-313.

Cheshier, S.C. (2010) *The New Class in Vietnam.* Doctoral Book. Queen Mary University of London, School of Business and Management.

Cleaver, F. (1999) "Paradoxes of participation: questioning participatory approaches to development". In: *Journal of international development* 11 (4), 597-612.

Cleaver, F. (2012) *Development through bricolage: rethinking institutions for natural resource management.* Routledge. 224.

Corbin, J. and Strauss, A. (2008) *Basics of qualitative research: Techniques and procedures for developing grounded theory.* SAGE Publications. 400.

Diglio, S. (1998) "The Changing Role of the State in Vietnam's Agricultural Policy". In: *Vietnam's Socio-Economic Development* 15, 39-46.

Ehlert, J. (2011) *Living with Flood. Local Knowledge in the Mekong Delta, Vietnam.* Doctoral Book. Rheinische Friedrich-Wilhelms-Universität Bonn, Department of Philosophy. 234.

Evers, H.-D. and Benedikter, S. (2009a) "Hydraulic Bureaucracy in a Modern Hydraulic Society – Strategic Group Formation in the Mekong Delta". In: *Water Alternatives* 2 (3), 416-439.

Evers, H.-D. and Benedikter, S. (2009b) *Strategic Group Formation in the Mekong Delta - The Development of a Modern Hydraulic Society*. ZEF Working Paper Series *Nr. 35*. Bonn: University of Bonn. http://www.zef.de/fileadmin/webfiles/downloads/zef_wp/WP35_Evers-Benedikter.pdf. (27.08.09).

Facon, T. (2007) "Performance of irrigation and participatory irrigation management: Lessons from FAO's irrigation modernization program in Asia". Paper presented at *The 4th Asian regional conference & 10th International seminar on Participatory irrigation management,* Tehran, Iran. 20.

Fforde, A. (2008a) "Political Processes in a Leninist State: The Politics of Rural Development Policy in contemporary Vietnam". Paper presented at *EUROVIET Conference,* Hamburg, 05.06.2008.

Fforde, A. (2008b) "Vietnam´s Informal Farmers´ Groups: Narratives and Policy Implications". In: *Südostasien Aktuell* 2008 (1), 3-36.

Fforde, A. (2011) "Contemporary Vietnam: Political Opportunities, Conservation Formal Politics, and Patterns of Radical Change". In: *Asian Politics & Policy* 3 (2), 165-184.

Fforde, A. and Nguyen Dinh Huan (2001) *Vietnamese Farmer´s Organisations: Economic and Political Rationales*. Adelaide. http://www.aduki.com.au/Farmers'%20Organisations%20-%20Final.pdf. (08.06.2011).

Fforde, A.J. (2010) "Rethinking the Political Economy of Conservative Transition: The Case of Vietnam". In: *Journal of Communist Studies and Transition Politics* 26 (1), 126-146.

Fontenelle, J.-P. (2000) "Water Management Decentralization in the Red River Delta, Vietnam. An Uncompleted Transition Process Toward Local Governance". Paper presented at *Eighth Biennial IASCP Conference 31/05-4/06/2000,* Bloomington. URL: http://www.inderscience.com/offer.php?id=2073. (Last accessed 22.11.2007).

Franks, T. and Cleaver, F. (2007) "Water Governance and Poverty: A Framework for Analysis". In: *Progress in Development Studies* 7 (4), 291–306.

Fritzen, S. (2002) "The 'Foundation of Public Administration'? Decentralization and its Discontents in Transitional Vietnam". Paper presented at *Asia Conference on Governance in Asia: Culture, Ethics, Institutional Reform and Policy Change,* Hong Kong. URL: http://kula.student.usp.ac.fj/class-shares/MG452/Articles%20&%20Conference%20Papers/Scott.pdf. (Last accessed 14.05.2008).

Fritzen, S. (2006) "Probing System Limits: Decentralisation and Local Political Accountability in Vietnam". In: *The Asia Pacific Journal of Public Administration* 28 (1), 1-23.

Fuhrmann, E. (2008) *Regulating Water in the Mekong Delta*. Diploma Book. Friedrich-Wilhelm Universität, South East Asian Studies.

Gainsborough, M. (2005) *Rethinking Vietnamese Politics: Will the Real State Please Stand Up? Bristol-Mekong Project Working Papers*. Bristol: University of Bristol. URL: http://www.bris.ac.uk/politics/grc/bvp/bvpworkingpapers/ (Last accessed: 07.11.2007).

Gainsborough, M. (2009) *Privatisation as State Advance: Private Indirect Government in Vietnam*. Bristol.

Gainsborough, M. (2010) *Vietnam. Rethinking the State*. New York, Chiang Mai: Zed Books Ltd, Silkworm Books. 224.

Garschagen, M.; Renaud, F.G.; Birkmann, J. (2011) "Dynamic Resilience of Peri-Urban Agriculturalists in the Mekong Delta Under Pressures of Socio-Economic Transformation and Climate Change". In: M. Steward; P. Coclanis (Eds.): *Environmental Change and Agricultural Sustainability in the Mekong Delta*. Heidelberg, London, New York: Springer Science and Business Media. 141-163.

Göbel, C. (2011) "Authoritarian consolidation". In: *European political science* 10 (2), 176-190.

Grossheim, M. (2004) "Village Government in Pre-colonial and Colonial Vietnam". In: B.J.T. Kerkvliet; D.G. Marr (Eds.): *Beyond Hanoi. Local Government in Vietnam*. Singapore: Institute of Southeast Asian Studies. 54-89.

GSO (2004) *Migration census of Vietnam 2004*. Ha Noi: General Statistics Office. http://www.gso.gov.vn/default.aspx?tabid=407&idmid=4&ItemID=3853. (05.11.2014).

GSO (2011) *Statistical Yearbook of Vietnam 2011*. Hanoi: General Statistics Office. http://www.gso.gov.vn/default_en.aspx?tabid=515&idmid=5&ItemID=12576. (05.11.2014).

GSO (2012) *Statistical Yearbook of Vietnam 2012*. Hanoi: General Statistics Office. http://www.gso.gov.vn/default_en.aspx?tabid=515&idmid=5&ItemID=12576. (05.11.2014).

Hammersley, M. and Atkinson, P. (2007) *Ethnography: Principles in practice*. Routledge. 278.

Hannah, J. (2007) *Local Non-Government Organizations in Vietnam: Development, Civil Society and State-society Relations*. Seattle: University of Washington.

Heberer, T. (2005) "Bottom-up Privatization under the Conditions of a "Vietnamese" Way to Socialism". In: G. Mutz; R. Klump (Eds.): *Modernization and Social Transformation in Vietnam: Social Capital Formation and Institution Building*. Hamburg: Institut für Asienkunde. 209-232.

Heinrich Böll Foundation (Ed.) (2005) *Towards Good Society - Civil Society Actors, the State, and the Business Class in Southeast Asia – Facilitators of or Impediments to a Strong, Democratic, and Fair Society?* Berlin: Heinrich Böll Foundation.

Hickey, G.C. (1968) *Village in Vietnam*. Yale University Press. 325.

Hicks, N. (2005) *Organisational Adventures in District Government: Central Control Versus Local Initiative in Long An Province, Vietnam*. Doctoral Book. Australian National University. 318.

Howie, C.A. (2011) *Co-operation and contestation: farmer-state relations in agricultural transformation, An Giang Province, Vietnam*. Doctoral Book. Royal Holloway, University of London, Centre for Developing Areas Research, Department of Geography.

Huynh Truong Huy and Le Nguyen Doan Khoi (2011) "Analysis of labour migration flows in the Mekong Delta of Vietnam". In: M.A. Stewart; P.A. Coclanis (Eds.): *Environmental Change and Agricultural Sustainability in the Mekong Delta*. Springer. 115-140.

Hy V. Luong (2003) "Introduction. Postwar Vietnamese Society: An Overview of Transformational Dynamics". In: H.V. Luong (Ed.): *Postwar Vietnam: Dynamics of a Transforming Society*. Oxford: Rowman & Littlefield. 1-26.

Hy V. Luong (2005) "The State, Local Associations, and Alternate Civilities in Rural Northern Vietnam". In: R.P. Weller (Ed.): *Civil Life, Globalization, and Political Change in Asia. Organizing between Family and State*. London and New York: Routledge. 123-147.

Jamieson, N.L. (1995) *Understanding Vietnam*. Berkeley: University of California Press. 428.

Joint Donor Report (2010) *Vietnam Development Report 2010. Modern Institutions*. Hanoi: Vietnam Development Information Center. (Last accessed: 26.01.2010).

Käkönen, M. (2008) "Mekong Delta at the Crossroad: More Control or Adaptation?". In: *Ambio: A Journal of the Human Environment* 37 (3), 205-212.

Keating, M. (2010) "Rethinking Governance: The Centrality of the State in Modern Society by Stephen Bell and Andrew Hindmoor". In: *Australian Journal of Public Administration* 69 (1), 103-106.

Kerkvliet, B.J.T. (2001a) "An Approach for Analysing State-Society Relations in Vietnam". In: *Journal of Social Issues in Southeast Asia (SOJOURN)* Vol. 16, No. 2, 238-278.

Kerkvliet, B.J.T. (2001b) "Introduction: Analysing the State in Vietnam". In: *Journal of Social Issues in Southeast Asia (SOJOURN)* 16 (2), 179-186. (Last accessed 20.07.2011).

Kerkvliet, B.J.T. (2003) "Authorities and the People: An Analysis of State-Society Relations in Vietnam". In: H.V. Luong (Ed.): *Postwar Vietnam: Dynamics of a Transforming Society*. Oxford: Rowman & Littlefield. 27-53.

Kerkvliet, B.J.T. (2004) "Surveying Local Government and Authority in Contemporary Vietnam". In: K. Benedict L. Tria; D.G. Marr (Eds.): *Beyond Hanoi.Local Government in Vietnam*. Nordic Institute of Asian Studies. 1-27.

Kerkvliet, B.J.T. (2005) *The Power of Everyday Politics. How Vietnamese Peasants Transformed National Policy*. Ithaca: Cornell University Press.

Kerkvliet, B.J.T. and Marr, D.G. (Eds.) (2004) *Beyond Hanoi. Local Government in Vietnam*. Singapore: Institute of Southeast Asian Studies.

Koh, D. (2001) "The Politics of a Divided Party and Parkinson´s State in Vietnam". In: *Contemporary Southeast Asia* 23 (3), 533-551.

Kono, Y. (2001) "Canal Development and Intensification of Rice Cultivation in the Mekong Delta: A Case Study in the Cantho Province, Vietnam". In: *Southeast Asian Studies* 39 (1), 70-85. (Last accessed 20.04.2010).

Kummu, M. (2003) "The natural environment and historical water management of Angkor, Cambodia". Paper presented at *The Fifth World Archeological Congress,* Washington DC. URL: http://users.aalto.fi/~mkummu/publications/kummu_WAC_WashingtonDC_2003.pdf. (Last accessed 05.06.2013).

Larsen, R.K. (2011) "Reinventing rural development in Vietnam: Discursive constructions of grassroots democracy during the renovation reform". In: *Asia Pacific Viewpoint* 52 (3), 316-332. URL: http://www.viet-studies.info/viet-studies/GrassRoot_Democracy.pdf. (Last accessed 05.11.2014).

Le Anh Tuan; Chu Thai Hoanh; Miller, F.; Bach Tan Sinh (2007) "Floods and Salinity Management in the Mekong Delta, Vietnam". In: Tran Thanh Be; Bach Tan Sinh; F. Miller (Eds.): *Challenges to Sustainable Development in the Mekong Delta: Regional and National Policy Issues and Research Needs*. Bangkok: Sustainable Mekong Research Network (Sumernet). 15-68.

Le Sam (1996) *Thủy nông ở đồng bằng sông Cửu Long - [Agriculture irrigation in the Mekong Delta]*. Ho Chi Minh city: Agriculture Publisher.

Le Trong Phuong (2013) "Tracing the discourses on civil society in Vietnam". In: G. Waibel; J. Ehlert; H.N. Feuer (Eds.): *Southeast Asia and the Civil Society Gaze: Scoping a Contested Concept in Cambodia and Vietnam*. Bonn Routledge. 171.

Long, N. (Ed.) (1989) *Encounters at the interface. A perspective on social discontinuities in rural development.* Wageningen: Agriculture University Wageningen.

Long, N. (Ed.) (2001) *Development sociology: actor perspectives* London and New York: Routledge - Taylor & Francis Group.

Long, N. and Liu, J. (2009) "The Centrality of Actors and Interfaces in the Understanding of New Ruralities: A Chinese Case Study". In: *Journal of Current Chinese Affairs* 38 (4), 63-84. URL: http://hup.sub.uni-hamburg.de/giga/jcca/article/view/118/118; (Last accessed: 03.07.2011).

MARD (2009) *Chiến lược phát triển nông nghiệp nông thôn 2011-2020 [The strategies for agriculture and rural development from 2011 to 2020].* In: MARD. Ha Noi. 82.

Marr, D.G. (2004) "A Brief History of Local Government in Vietnam". In: B.J.T. Kerkvliet; D.G. Marr (Eds.): *Beyond Hanoi. Local Government in Vietnam*. Singapore: Institute of Southeast Asian Studies. 28-53.

Mart A. Stewart and Coclanis, P.A. (2011) *Environmental Change and Agricultural Sustainability in the Mekong Delta.* Advances in Global Change Research 45. London, New York: Springer. 455.

Meinzen-Dick, R.S. and Nkonya, L.K. (2007) *Understanding legal pluralism in water and land rights: lessons from Africa and Asia.* CABI London.

Miles, M.B. and Huberman, A.M. (1994) *Qualitative data analysis: An expanded sourcebook.* Sage Publications, Inc. 352.

Molle, F.; Mollinga, P.P.; Wester, P. (2009) "Hydraulic bureaucracies and the hydraulic mission: flows of water, flows of power". In: *Water Alternatives* 2 (3), 328-349.

Mollinga, P.P. (2008) "Water, Politics and Development: Framing a Political Sociology of Water Resources Management". In: *Water Alternatives* 1 (1), 7-23.

Neuman, W.L. (2006) *Social research methods: Qualitative and quantitative approaches.* 6th edition. Boston: Pearson. 600.

Nguyen Anh Thuy (2011) *Tài liệu hỏi đáp: Xây dựng nông thôn mới cấp xã (phục vụ cho công tác tuyên truyền trong nhân dân) [Booklet: building the new rural area at commune level (serving the mobilisation in the community)].* Can Tho city: Leading board for national target programme about constructing new rural area in Can Tho city

Nguyen Duy Can; Le Thanh Duong; Nguyen Van Sanh; Miller, F. (2007) "Livelihoods and Resource Use Strategies of Farmers in the Mekong Delta". In: Tran Than Be; Bach Tan Sinh; F. Miller (Eds.): *Challenges to sustainable Development in the Mekong Delta: Regional and National Policy Issues and Research Needs.* Bangkok: The Sustainable Mekong Research Network (sumernet). 69-98.

Nguyen Ngoc De (2006) *Farmers, Agriculture and Rural Development in the Mekong Delta of Vietnam*. Doctoral Book. Mie University, Graduate School of Bio-resources.

Nguyen The Thang (2012) "Tập thể lãnh đạo cá nhân phụ trách theo tư tưởng Hồ Chí Minh ". In: *Tạp chí cộng sản* 68, 7-11. URL: http://tutuonghochiminh.vn/cmvn/tap-the-lanh-dao-ca-nhan-phu-trach-theo-tu-tuong-ho-chi-minh.d-842.aspx. (Last accessed: 04.11.2014).

Nguyen Thi Hien (2008) "Irrigation fees from the perspective of socio-economics". In: Nguyen Xuan Tiep (Ed.): *Participatory Irrigation Management and Emerging Issues - Nông dân tham gia quản lý công trình thủy lợi và những vấn đề đang đặt ra* Ha Noi: The Agricultural Publishing House. 148-156.

Nguyen Thi Phuong Loan (2010a) *Legal Framework of the Water Sector in Vietnam*. ZEF Working Paper No. 52. Bonn: University of Bonn. http://www.zef.de/fileadmin/webfiles/downloads/zef_wp/WP52_Nguyen.pdf. (Last accessed: 23.08.2010).

Nguyen Thi Phuong Loan (2010b) *Problems of Law Enforcement in Vietnam: The Case of Wastewater Management in Can Tho City*. ZEF Working Paper No. 53. Bonn: University of Bonn. http://www.zef.de/fileadmin/webfiles/downloads/zef_wp/WP53_Nguyen.pdf. (Last accessed: 23.08.2010).

Nguyen Trung Vinh and Nguyen Khuong Ba (Eds.) (2008) *Lịch sử phong trào nông dân và hội nông dân tỉnh Cần Thơ 1930-2008 [The history of farmer campaign and the Farmer´s Union of Can Tho province 1930-2008]*. Can Tho: Can Tho´s Farmer´s Union.

Nguyen Van Sanh; Vo Tong Xuan; Tran An Phong (1998) "History and Future of Farming Systems in the Mekong Delta". In: V.T. Xuan;; S. Matsui; (Eds.): *Development of Farming Systems in the Mekong Delta of Vietnam*. Ho Chi Minh city: Ho Chi Minh City Publishing House.

Nguyen Xuan Tiep (2008a) "Participatory Irrigation management - PIM". In: Nguyen Xuan Tiep (Ed.): *Participatory irrigation management and emerging issues - Nông dân tham gia quản lý công trình thủy lợi và những vấn đề đang đặt ra*. Ha Noi: The agricultural pulishing house. 22-42.

Nguyen Xuan Tiep (Ed.) (2008b) *Participatory Irrigation Management and Emerging Issues - Nông dân tham gia quản lý công trình thủy lợi và những vấn đề đang đặt ra*. Ha Noi: The Agricultural Publishing House.

Norlund, I.; Dang Ngoc Dinh; Bach Tan Sinh; Chu Dung; Dang Ngoc Quang; Do Bich Diem; Nguyen Manh Cuong; Tang The Cuong; Vu Chi Mai (2006) *The Emerging Civil Society. An Initial Assessment of Civil Society in*

Vietnam. Hanoi: Vietnam Institute of Development Studies (VIDS), UNDP Vietnam, SNV Vietnam and CIVICUS Civil Society Index. http://www.civicus.org/new/media/CSI_Vietnam_report%20.pdf. (Last accessed: 11.08.2008).

Office for Social Evaluation and Consultancy - OSEC (2011) *Research report from rural labour and employment in Vietnam*. Ha Noi: Vietnam union of Science and technology Associations - VUSTA. http://www.ilo.org/wcmsp5/groups/public/---asia/---ro-bangkok/---ilo-hanoi/documents/publication/wcms_171760.pdf. (Last accessed: 25.09.2012).

Ooi, K.G. (2004) *Southeast Asia: A Historical Encyclopedia, from Angkor Wat to Timor. RZ. volume three*. 1. Abc-clio.

Ostrom, E. (1990) *Governing the commons: The evolution of institutions for collective action*. Cambridge: Cambridge University Press. 298.

Ostrom, E. (1999) "Social capital: a fad or a fundamental concept". In: P. Dasgupta; I. Seraeldin (Eds.): *Social capital: A multifaceted perspective*. Washington, DC: The World Bank. 172-214.

Pham Anh Tuan and Shannon, K. (2011) "Water Management in Vietnam Indigenous Knowledge and International Practices: The Case of the Red River Delta". Paper presented at *N-AERUS XI Urban Knowledge Cities of the South* Brussels URL: http://www.n-aerus.net/web/sat/workshops/2010/pdf/PAPER_pham_t.pdf. (Last accessed: 04.11.2014).

Pham Cong Huu (2011) *Planning and Implementation of the Dyke Systems in the Mekong Delta, Vietnam*. Doctoral Book. Rheinische Friedrich-Wilhelms-Universität Bonn, Department of Mathematics and Science. 181.

Pham Cong Huu (2012) *Planning and Implementation of the dyke systems in The Mekong Delta*. Berlin, Münster, Wien, Zürich, London: Lit Verlag.

Pham Quang Tu; Đao Trong Hung; Tran Huu Hiep; Ho Thanh Son (2013) *Phân tích các bên liên quan trong quản lý tài nguyên nước và thích ứng biến đổi khí hậu vùng đồng bằng sông Cửu Long - [Assessing the aspects of water resources management and the ability to adapt to climate change in the Mekong Delta]*. 75.

Pham Van Lam (...) *Cơ sở khoa học của các giải pháp phòng chống dịch rầy nâu, bệnh virus lúa cỏ và lúa lùn xoắn lá - [Scientific proof of the techniques to prevent and cure plant hopper epidemic, virus diseases]*. http://opac.lrc.ctu.edu.vn/opac/. (Last accessed: 11.09.2013).

Popkin, S.L. (1979) *The rational peasant: The political economy of rural society in Vietnam*. University of California Press. 332.

Porter, G. (1993) *Vietnam. The Politics of Bureaucratic Socialism.* Politics and International Relations of Southeast Asia. Ithaca, London: Cornell University Press. 227.

Rama, M. (2008) *Making Difficult Choices: Vietnam in Transition - based on conversations with H.E. Vo Van Kiet, with Professor Dang Phong and Doan Hong Quang.* Commission on Growth and Development. Washington DC: World Bank.

Rambo, A.T. (2005) *Searching for Vietnam: Selected writings on Vietnamese culture and society.* Kyoto Area Studies on Asia 9. Trans Pacific Press. 456.

Reis, N. (2010) *Tracing and Making the State. Policy Practices and Domestic Water Supply in the Mekong Delta, Vietnam.* Doctoral Book. Rheinische Friedrich-Wilhelms-Universität Bonn, Department of Agriculture. 240.

Reis, N. (2013) "Civil society and political culture in Vietnam". In: G. Waibel; J. Ehlert; H.N. Feuer (Eds.): *Southeast Asia and the Civil Society Gaze: Scoping a Contested Concept in Cambodia and Vietnam.* Bonn: Routledge. 77.

Renault, D.; Facon, T.; Wahaj, R. (2007) *Modernizing irrigation management – the MASSCOTE approach.* FAO Irrigation and Drainage Paper *Nr. 63.* Rome FAO.

Rosen, L. (1984) *Bargaining for reality: The construction of social relations in a Muslim community.* Chicago: University of Chicago Press. 217.

Rudestam, K.E. and Newton, R.R. (2007) *Surviving your book: A comprehensive guide to content and process.* California: SAGE Publications, Inc. 328.

Sansom, R.L. (1969) "The motor pump: a case study of innovation and development". In: *Oxford Economic Papers*, 109-121.

Scott, J.C. (1987) *Weapons of the Weak: Everyday Forms of Peasant Resistance.* Yale University Press. 392.

Scott, J.C. (1998) *Seeing like a state: How certain schemes to improve the human condition have failed.* Yale University Press. 464.

Sebesvari, Z.; Huong, L.T.T.; Toan, P.V.; Arnold, U.; Renaud, F.G. (2012) "Agriculture and Water Quality in the Vietnamese Mekong Delta". In: F. Renaud; C. Künzer (Eds.): *The Mekong Delta System - Interdisciplinary Analyses of a River Delta.* Heidelberg: Springer. 331-361.

Son Nam (1997) *Lịch Sử Khẩn Hoang Miền Nam - [The history of land reclaimation of the South].* Ho Chi Minh city: Nhà xuất bản Trẻ - [Youth Publisher].

Son Nam (2003) *Son Nam: Nói về miền Nam, Cá tính miền nam, Thuần phong mỹ tục miền nam [Son Nam: Talk about the South, Th r South personality, The Southern traditions].* Ho Chi Minh: Youth Publishing House. 404.

Southern Institute For Water Resources Planning - SIWRP (2011a) *Quy hoạch tổng thể thủy lợi đồng bằng sông Cửu Long trong điều kiện biến đổi khí hậu-nước biển dâng - [The general planning for irrigation in the Mekong Delta in the condition of climate change and sea water rise]* Ho Chi Minh city Southern Institute For Water Resources Planning - SIWRP.

Southern Institute For Water Resources Planning - SIWRP (2011b) *Vietnam-Netherlands Mekong Delta Masterplan project: Draft Report on water resources infrastructure assessment study in the Mekong Delta.* Ho Chi Minh city Southern Institute For Water Resources Planning - SIWRP.

Sox, D.G. (1972) *Resource-use systems of ancient Champa* Hawaii: University of Hawaii. 418.

Statistical office of Can Tho city (2008) *Statistical Yearbook of Can Tho city 2007*. Can Tho: Statistical Office of Can Tho city.

Statistical office of Can Tho city (2011) *Statistical Yearbook of Can Tho city 2010*. Can Tho: Statistical Office of Can Tho city.

Sub-Department of Plant Protection Can Tho city (2011) Some highlighted issues in directing the production of winter-spring rice season 2011-2012. No. 404/CC.BVTV.

SWIRP (2012) Phê duyệt Quy hoạch thủy lợi đồng bằng sông Cửu Long giai đoạn 2012 - 2020 và định hướng đến năm 2050 trong điều kiện biến đổi khí hậu, nước biển dâng - [Approval of the irrigation planning for the Mekong Delta in 2012-2020 and orientation to 2050 in the condition of climate change and sea water rise]. http://siwrp.org.vn/?id_pnewsv=490&lg=vn&start=0. (Last accessed: 04.11.2014).

To Van Truong (2011) Đê bao, bờ bao ở đồng bằng sông Cửu Long [Dykes and embarkments in the Mekong Delta]. http://www.cpo.vn/de-bao-bo-bao-o-dong-bang-song-cuu-long_n1616.aspx. (Last accessed: 06.08.2013).

Truong Dong Loc (2006) *Equitisation and Stock-Market Development, The Case of Vietnam*. Doctoral book. University of Groningen. 258.

Uphoff, N.; Wickramasinghe, M.; Wijayaratna, C. (1990) "" Optimum" Participation in Irrigation Management: Issues and Evidence from Sri Lanka". In: *Human Organization* 49 (1), 26-40.

Vo Khac Tri (2012) "Hydrology and Hydraulic Infrastructure systems in the Mekong Delta, Vietnam". In: F.G. Renaud; C. Kuenzer (Eds.): *The Mekong Delta System. Interdisciplinary Analyses of a River Delta*. Heidelberg: Springer. 50-81.

Vo Thi Guong and Nguyen My Hoa (2012) "Aquaculture and Agricultural Production in the Mekong Delta and its Effects on Nutrient Pollution of Soil and Water". In: F. Renaud; C. Künzer (Eds.): *The Mekong Delta System - Interdisciplinary Analyses of a River Delta*. Heidelberg: Springer. 363-393.

Waibel, G. (2010) *State Management in Transition: Understanding Water Resources Management in Vietnam*. ZEF Working Paper No. 55. Bonn: University of Bonn. http://www.zef.de/fileadmin/webfiles/downloads/zef_wp/WP55_Waibel.pdf. (Last accessed: 22.04.2010).

Waibel, G. (2013) "Grasping discourses, researching practices". In: G. Waibel; J. Ehlert; H.N. Feuer (Eds.): *Southeast Asia and the Civil Society Gaze: Scoping a Contested Concept in Cambodia and Vietnam*. Bonn: Routledge. 1-.

Waibel, G. and Benedikter, S. (2013) *Central Directives, Local Administration and the Formation of Water User Groups in the Mekong Delta, Vietnam*. Unpublished Work.

Waibel, G.; Benedikter, S.; Reis, N.; Genschick, S.; Loan Nguyen; Pham Cong Huu; Tran Thanh Be (2012) "Water Governance Under Renovation? Concepts and Practices of IWRM in the Mekong Delta, Vietnam". In: F.G. Renaud; C. Kuenzer (Eds.): *The Mekong Delta System. Interdisciplinary Analyses of a River Delta*. Springer. 167-198.

Wescott, C.G. (2003) "Hierarchies, networks and local government in Viet Nam". In: *International Public Management Review* 4 (2), 20-40.

Wischermann, J. (2013) *Civic organizations in Vietnam's one-party state: Supporters of authoritarian rule?* GIGA working paper *Nr. 228*. German Institute of Global and Area Studies. http://www.giga-hamburg.de/en/system/files/publications/wp228_wischermann.pdf. (Last accessed: 05.11.2014).

Wischermann, J. and Nguyen Quang Vinh (2003) "The Relationship Between Civic and Governmental Organizations in Vietnam: Selected Findings". In: B.J.T. Kerkvliet; D.W.H. Koh; R.H.-k. Heng (Eds.): *Getting Organized in Vietnam. Moving in and around the Socialist State*. Singapore: Institute of Southeast Asian Studies. 185-233.

Yin, R.K. (2003) *Case Study Research: Design and Methods.* SAGE Publications, Inc. 312.

Zheng, Y. and Lye, L.F. (2011) "Is Democratization Compatible with China's One-Party System?". In: *Political Parties, Party Systems and Democratization in East Asia*, 17.

Annex

PLAN OF INQUIRY: LOCAL IRRIGATION MANAGEMENT IN THE MEKONG DELTA

Most researches about the Delta's water management have strongly focussed on the domain of state management but have often stopped short of analysing the level of practise. The accounts of the Southern villages and the local interactions in guiding agriculture irrigation management are limited to a few works. For instance, Hickey (1968), in a research on the Southern villages rare at the time, asserts the influence of Cham and Khmer techniques on the current practise of rice farming and the open-mindedness of the villagers in the Delta in guarding the old ways while adopting the new arrangements; Jamieson (1995) explains Vietnamese culture, social organisation and behaviours during wartime based on the balance between the *Yin* of flexibility, feeling, empathy, and spontaneity, and the *Yang* of rigid hierarchy and redundancy. In more recent work on the Delta communities, Hicks (2005) contributes to the understanding of the state-society relationship in Vietnam through the state-in-society approach, and with a focus on the district level, the author took note of the central state acceptance of local experiments during the reform while keeping its influence and interests felt at the district level. The three researches exploring the Delta's context in the previous periods indeed require an update on the interaction between the state and local actors. In particular, a contemporary empirical account of the state-society relationship covering the rapidly changing conditions due to private farming and growing marketisation in the Delta is missing. Recently, Howie (2011), by researching the pumping groups in An Giang, confirmed the important role of local actors and the 'dialogic' interaction between the authorities and the farmers, and how farmers co-operate to overcome the common threat, and that farmers make choices through social cohesion while individual wishes are not totally suppressed. However, Howie's work, while praised for its empirical contributions, didn't provide the answer for how the state and its policies interacts with local arrangements, which requires more evidence and a comprehensive explanation of individual behaviours deciding the local practises. The main motivation for my research was to advance our understanding of the processes that guide and decide the local practises of irrigation management in the Mekong Delta. This will, in turn, contribute to an analysis of the Vietnamese state and its role and status in various contemporary contexts. Therefore, the research has three purposes: to explore the behaviours of present actors involved in irrigation activities, to describe their interactions and negotiations with other parties, and to conceptualise the local dynamic within the realm of institutions and the state-society relationship. By studying the Mekong Delta in this manner, this research sought to answer the following three exploratory questions:

- Who is involved in local irrigation management in the Mekong Delta?
- How and why are things happening at the local level, and what contextual forces guide irrigation practise in the contemporary Mekong Delta?
- What is the stance of the state vis-a-vis guiding the management of irrigation practise?

To frame these questions theoretically, I chose an abductive strategy common in political ecology, which "*begins by describing these activities and meanings and then deriving from them categories and concepts that can form the basis of an understanding of the problem at hand*" (Blaikie 2010:89). With abductive strategy, the research focusses on the meaning and interpretations, the motivations and intentions that people use in their everyday lives as a means to discover why people eventually do what they do; these understandings are then elevated into concepts and theories (ibid). Gaining the data necessary for an abductive approach generally requires anthropological methods and a fairly limited case study that allows for considerable time in the field. My general timeline thus fell into the following steps:

1. Selecting a case and managing field access: here, the WISDOM project team already had research access and topics to choose from.
2. Data collection: using ethnographic methods to research individual farmers, the settings of farm-relevant events and decisions, state actors and other organisational figures.
3. Data reduction and analysis: to abstract or generate technical concepts from the lay patterns.

In searching for the understanding of local interactions and negotiations of irrigation practises which then clarifies the contemporary state-society interaction in agriculture and irrigation management, this research is conceptualised with the theme of institutions and state power dimensions (explained in detail in chapter 2). In the process of analysing the irrigation institutions at the "bureaucratic-informal" interface, the bargaining of individual behaviour and the negotiations between stakeholders are analysed. Such analysis requires a detailed empirical account of the real-life context. Amongst others, the ethnographic case study method appears to be the best choice.

To conduct this research, I assembled various strategies for case study research used in settings similar to mine in some way (e.g., Eckstein 1975:85; Jervis 1990; Ragin 1992, King, Keohane and Verba 1994; Yin 1994; George and Bennett 2005). The case study is especially important in this context because the research is trying to build an understanding of the local practises that is very much dependant on its context of rural farming in the Mekong Delta in the present period. As defined by Yin (2003:13), "*a case study is an empirical enquiry that investigates a contemporary phenomenon within its real-life context, especially when the boundaries between phenomenon and context are not clearly evident.*" This consideration is important in the Mekong Delta, as we are examining evolving practises that cannot be 'pinned down' by surveys or other static instruments; in this setting, a case study is also more suitable than an historical approach or experimental research (natural experiments) (Yin 2003:8). Within case studies, ethnographic methods involve direct observations, interviews, limited participation, and documentation of events. To manage the inevitable contradictions that emerge in research areas such as Vietnam, I also applied a triangulation approach aimed at acquiring comparable data from multiple methods and sources. The amount and adequacy of data collected tell that case studies and ethnographic methods are best for doing this research. The effectiveness of this approach emerged at many different stages of the research.

Stage 1: Choosing the cases and entering the field

The research started with gathering basic information about the irrigation sector and its management in the Mekong Delta from previous researches and fieldwork accounts, as well as through expert interviews. With this background information, I decided to start the research from the top of the state hierarchy, working down to the provincial to commune and hamlet levels.

Although this research aims to understand the local practises, especially in Vietnam, it is necessary to legitimate the researcher's presence in the field. The researcher's identity had been set up through a bureaucratic procedure for researchers working for a foreign institute. I was first authorised to be an affiliate researcher under the Department of Foreign Affairs, Can Tho University. This is different from other foreign researchers in the same WISDOM project in that their monthly work-plans needed to be approved by Can Tho's Department of Foreign Affairs before the activities were conducted; as a Vietnamese, I only needed an introduction letter from Can Tho University (Figure A-18) to get an appointment with the provincial state agencies of Can Tho, An Giang and Bac Lieu. Again, with the advantage of being Vietnamese, I could move down to district- and commune-level agencies with the same letter shown, or without any letter. Appointments at the district and commune were made by phone in advance, and often started from the highest government management office (the People's Committee) from which I could get permission, recommendations and phone numbers to contact the other offices for interviews or data granting. As a researcher in an international project, official procedures are essential and important to open the gates to state agencies and research sites. The most important purpose is to clarify the researcher's identity in the field, thus avoiding any possible legal problems.

By building networks and remaining in contact with local cadres (communes and hamlets), I could continually clarify my status as a researcher to the farmers and, importantly, I was readily granted access for the first observations and interviews. In the local areas in Vietnam, the commune and hamlet leaders are the gatekeepers to the field. Starting from the state system fulfils another goal of this research, namely the task of analysing the procedures and protocols of state management in the irrigation sector. While there might be the possibility of negative effects (i.e., suspicion) if farmers have poor relations with local leaders, I found that this was not an issue because the topics are not socially or politically sensitive, and many local cadres are also social leaders due to their prestige and local likability.

The data collection began with visits to the provincial state agencies including the Department of Agriculture and Rural Development, the Sub-department of Irrigation, the Farmers' Union, the Association of Cooperatives of Can Tho, An Giang, Bac Lieu, Sub-department of Rural Development of Can Tho and An Giang. The aims were to identify the current status of the irrigation infrastructure, the mode of management for agriculture and irrigation, the task division between agencies, the overall information about groups and co-operatives in the province, and the contact information of district state agencies. With the availability of data through previous work in the WISDOM project, the better connections as well as the convenience to conduct the field visits, Can Tho was chosen as the main case area. The four districts of Can Tho were initially surveyed for suitability, including Thoi Lai, Co Do, Vinh Thanh and Thot Not (Figure A-17). The topics of conversation at the district level were similar to those at the provincial level, with adjustments for scale. Information about the organisation of irrigation activities was obtained from 24 communes based on visits to people's committee offices (e.g., visiting the commune, Figure A-19). After gaining an overall picture of the district management structure for irrigation management, two activities were chosen as case studies: field drainage and canal dredging. The two activities have the same feature of creating the interface for interaction and negotiation among stakeholders, both in state-designed meetings and everyday negotiation.

Thoi Lai district is located in the middle of Can Tho and is one of the rural districts dominated by agricultural activities. The district belongs to the Trans-Bassac depression zone (see Figure 3-4 in chapter 3 for the eco-hydrological zones of the Mekong Delta), average flood area 0,6-2m (Pham Cong Huu 2011:45-46). The district can be divided into two areas. One is the area that is close to the Hau River or the Bassac River, including Truong Thanh, Dinh Mon and Xuan Thang communes. These communes are characterised by natural canals (canals which were constructed based on the natural waterways) and dominated by small holdings and/or separate pieces of land. Because small land-holding is often associated with poor households, which restricts the owners from the incentive to sacrifice own benefit for the common one, farmers in these communes are more likely to adopt individualistic behaviours in farming activities. The second part of the district, bordering with Kien Giang province, includes communes with bigger land-holding and constructed canals (new canals which were created later from French colonial time until the period after 1975). Among the communes in this area, Dong Binh and Dong Thuan are the dominant agricultural production communes with high-quality rice, a nearly completed stable dyke system, and active co-operative groups. Because of the diversity in land-holding, the possible access and the differences in farmer group characteristics, this district was selected as the case area. In the Thoi Lai district, the organisation and management of the two activities were explored in three communes: Xuan Thang, Truong Xuan B, and Dong Thuan (Figure A-17), making up the three case studies. The basic characteristics of the communes are described in Table A-8 below. The three cases are supposed to complement each other in identifying the patterns of interaction and negotiation in the local practises.

Figure A-17: The research communes (produced by Dunja Krause, 2015)

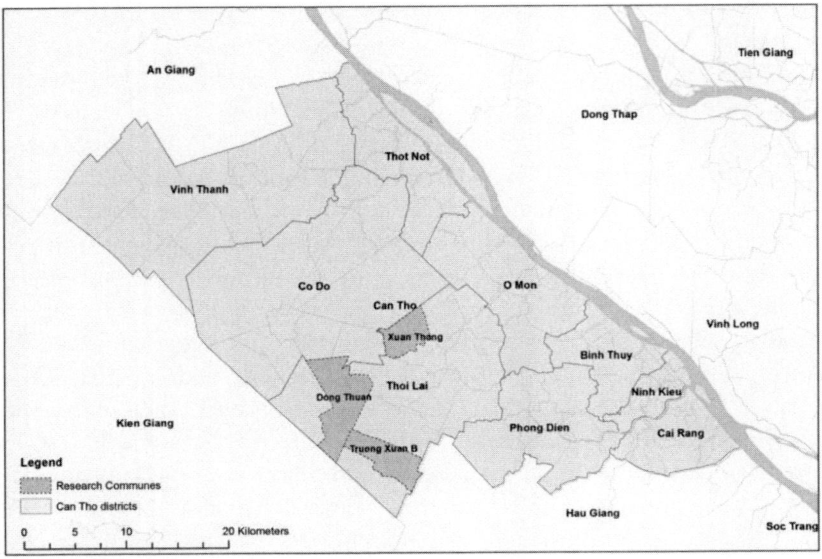

Table A-8: Characteristics of the selected case communes in Thoi Lai district

Commune	Majority of land-holding	Group situation - preliminary stage	Collective field drainage
Xuan Thang	Small (less than 1 hectare)	22 groups - only on paper	Private pumping - some collective pumping zones
Truong Xuan B	Medium to large (< 1 ha.-3 ha.)	No group	Collective pumping in zones
Dong Thuan	Large (>3 ha.)	Active groups	Collective pumping in zones

*Note: the categorization of land-holding is based on the estimation of the farmers and local cadres. Their perceptions of big and small land-holding do not necessarily follow any official category.

Stage 2: Data collection

There were three types of data collected for this research:

- Primary data from the case study of three communes, mostly from observations and interviews.
- Secondary data: public statistics of agriculture, data on irrigation infrastructure, management procedure, state documents, maps and diagrams of the hydraulic networks, demographic data, and so on.
- Tertiary data: literature on agriculture, irrigation, the state, and cooperative groups.

The data sources were diverse. Typical sites included offices, meetings, cafeterias, and rice fields, where many events occurred in natural social settings in which the researcher was present with minimal influence (state-designed meetings or farmers' gatherings). Data were also obtained from semi-natural settings (interviews) and social artefacts (public and private documents, photographs).

With the case studies chosen, field research generally followed ethnographic protocols (Hammersley and Atkinson 2007). In my case, the research combined observation and semi-structured interview methods, which was best tailored to attaining a 'systematic ethnographic understanding of social life' (Long 2001:15), and to provide comparable stories. Participant observations were focussed mostly on events and meetings, whereas semi-structured interviews (Figure A-21) were mostly conducted with individuals and small groups (husband and wife, or between neighbours) in the hamlets. With an assistant, I was present during daily arrangements of the field drainage activities, state-designated meetings (see pumping meetings in Figure A-20) and the more local-designated meetings (morning coffee drinking, or in the rice fields - see Picture 5-13 in chapter 5). During the observations and interviews, documents such as state production planning documents (Figure A-23), group decisions (see Figure 4-7, chapter 4), meeting minutes, pumping agreements (see Picture 5-9 in Chapter 5) and artefacts (Figure A-22) were collected. Together, the collected data illuminate how and why local negotiation of irrigation activities and the state-society interaction are happening in practises.

Most interviews included an ice-breaking component in which I shared my background, joked and told stories in order to build trust and encourage participants to open up. Confidentiality was agreed upon and promised by the researcher for certain sets of data. Participants´ identities are treated as private property and will not be mentioned in this work. Through careful watching and listening, data from observations and interviews were collected in the form of maps, diagrams, photographs, tape recordings, jotted field notes and detailed notes written away from the field. The detailed notes include transcriptions of jotted notes, direct observation notes, the researcher´s interpretation of the notes, and analytical memos. Interviews took between 30 and 120 minutes.

To choose interviews, I initially started with single-stage non-probability sampling, and later moved to purposive and snow-ball sampling. This initial sampling method helped to clarify the apparent objectivity of the researcher and to dispel any sense of arbitrariness. After the first interviews with the sampling method, some patterns and concepts began to emerge from the data that need to be further explored and explained, thus encouraging different types of sampling. For example, participants were selected follow-up interviews or selected based on observations at events (i.e., pumping meetings, commune meetings), which were selected purposively.

Data collection tapered off as described by Josselson and Lieblich (2003, cited in Rudestam and Newton 2007:108); when the results start to become redundant, the unique contribution of each informant grows marginal. This saturation point was reached after 10 months, with 52 offices visited, 67 individuals interviewed (once or more), 29 meetings observed, and many field walks.

Stage 3: Data reduction and analysis

Raw data in the form of field notes are reduced by open coding to breaking the data down into categories and sub-categories (Neuman 2006:461; Corbin and Strauss 2008:195). The categories are organised in cause-effect relation, making connections between categories to discover regularities, variations and singularities in the data. At this stage, it begins to construct or link to theories and themes – axial coding (Neuman 2006:462). The final stage is selective coding, in which I selected the data that will support the conceptual coding categories that were collected and conceptualised during axial coding. The first two steps could also be called pattern-matching and explanation-building, as stated by Miles and Huberman (1994:111). The analysis of data is researcher-specific as the classification is not a neutral process; in accordance with Neuman (2006:212), I do have a purpose in mind that provides boundaries for the analysis.

Limitations of this research

While fulfilling the task of collecting data to answer the research questions, there were a number of settings in which I had a clear disadvantage that limited some aspects of my research. One significant difficulty came from the poor access to certain types of events for direct observation. As a woman, for example, I was not able to participate in overnight pumping and thus could not directly record the negotiation, conflict resolution, and resolution of typical problems such as dyke damage. To some degree, this was remedied by visiting the following morning and getting a fresh account of the previous night. In this and other situations in which gender played a role, I could have sought the help of a male assistant. Besides the long-term difficulties of working with a male assistant, this was not considered to be an ideal solution to some of the dilemmas. For example, during night-pumping, there is the risk that a male assistant may be obliged to drink wine – something that female interviewers are often able to avoid without souring relations.

Another issue that signified some limitations to data collection surrounded various conflict cases. Often, it is not easy to access the details of a conflict because they remain contentious among the parties involved. Furthermore, this also led to missing interviews with difficult households (in both field drainage and canal dredging) and therefore to a deficit of explanations for the reasons for their actions (or inaction). If this research focussed solely on the conflict resolution and more fieldwork was allowed, it is more likely that a researcher could build rapport with difficult households and further exploration of the issues would be possible. However, within the framework of this research, interviews with difficult households present the risk of losing relationships and trust on one or the other side of the conflict, or even increasing tension among stakeholders because of the research activities.

A third challenge presented itself after returning from the field. When cross-checking or double-checking became necessary, it was clear that long-distance contact was only viable with a small group of state leaders at the provincial and district levels; I had insufficient protocols in place to communicate about specific data with farmers.

Beyond these more accumulative issues, notes are left throughout the text to help readers understand the context of the interviews and observations as well as any other limitations. Any additional ambiguities are the responsibility of the author.

Figure A-18: Example: official letter issued by Can Tho University to introduce the PhD researcher to visit state agencies

BỘ GIÁO DỤC VÀ ĐÀO TẠO
TRƯỜNG ĐẠI HỌC CẦN THƠ

CỘNG HÒA XÃ HỘI CHỦ NGHĨA VIỆT NAM
Độc lập – Tự do – Hạnh phúc

Số: 1053/ĐHCT-HTQT
V/v Giới thiệu Nghiên cứu sinh đến làm việc với đơn vị.

Cần Thơ, ngày 21 tháng 6. năm 2011

Kính gửi: - Chi Cục Thủy lợi tỉnh An Giang.

Trường Đại học Cần Thơ xin giới thiệu: Cô Huỳnh Thị Phương Linh

Là Nghiên cứu sinh thuộc dự án WISDOM giai đoạn II- Đại học Cần Thơ (Dự án hợp tác giữa Đức và Việt Nam về Hệ thống thông tin có liên quan đến nước để phát triển bền vững ĐBSCL).

Đến liên hệ công tác tại Chi Cục Thủy lợi tỉnh An Giang về các nội dung sau:

- Tìm hiểu thông tin tổng quát về các công trình thủy lợi trên địa bàn tỉnh.
- Tình hình vận hành, quản lý các công trình thủy lợi, các điểm thuận lợi và khó khăn hiện tại.
- Tổ chức hoạt động của Chi cục thủy lợi và các đơn vị trực thuộc ở cấp huyện, xã.
- Mối liên hệ với các cơ quan, đơn vị khác trong quản lý thủy lợi.
- Thông tin thứ cấp khác có liên quan như báo cáo, tờ rơi, bản đồ.

Thời gian: *từ 9:00 đến 11:00, ngày 19/7/2011.*

Trường Đại học Cần Thơ rất mong Chi Cục Thủy lợi tỉnh An Giang hỗ trợ và tạo điều kiện thuận lợi cho cô Huỳnh Thị Phương Linh tiếp cận và hoàn thành nội dung liên quan đến đề tài nghiên cứu như trên.

Trân trọng cám ơn.

KT.HIỆU TRƯỞNG
PHÓ HIỆU TRƯỞNG

TRƯỜNG ĐẠI HỌC CẦN THƠ

Lê Việt Dũng

Nơi nhận:
- Như trên;
- Lưu KHTH- HTQT.

Figure A-19: Visit to the commune (Hai Trieu, 15.08.2011)

Figure A-20: Observations

Production meetings at commune

Pumping meetings at hamlet

Figure A-21: Interviews with farmers

Figure A-22: Artefact examples – Top: Sketch of canal networks in the commune – Bottom: Record of fueling calculations of a group of 17 farmers

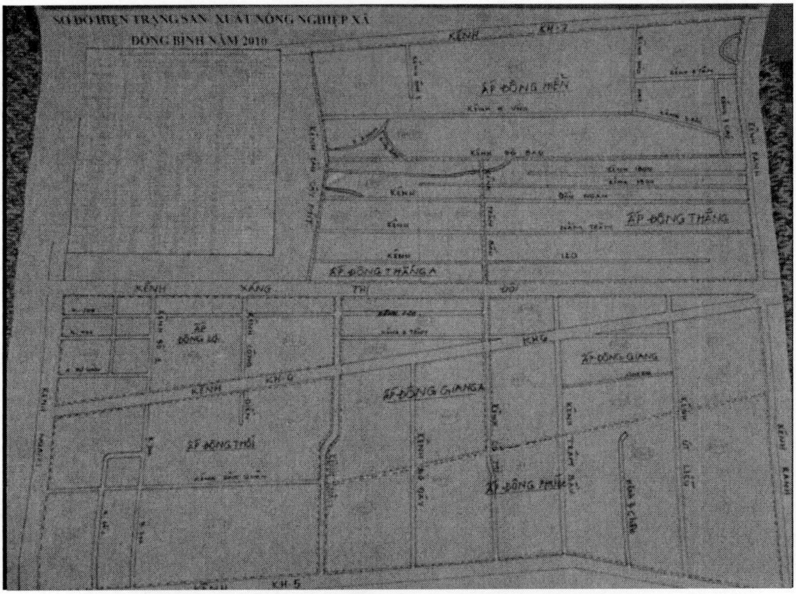

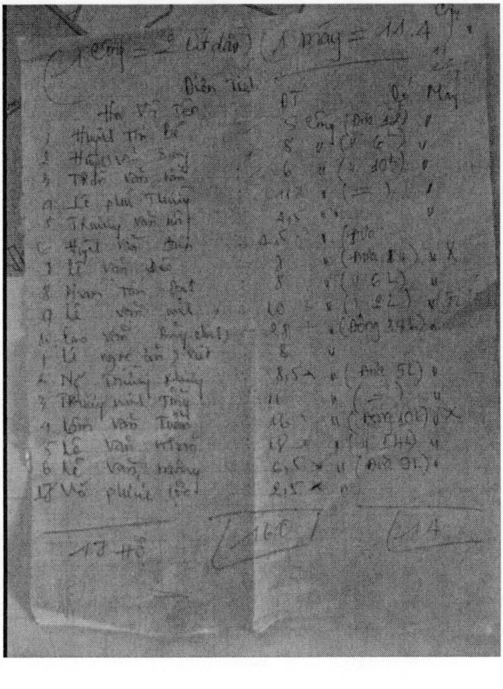

Figure A-23: Decision of the Commune´s People´s Committee about the establishment of the production board for the Winter-Spring season of years 2011-2012 at commune

ỦY BAN NHÂN DÂN
XÃ TRƯỜNG XUÂN B
* * *

CỘNG HÒA XÃ HỘI CHỦ NGHĨA VIỆT NAM
Độc lập - Tự do - Hạnh phúc

Số :199/QĐ-UBND

Trường Xuân B, ngày 28 tháng 10 năm 2011

QUYẾT ĐỊNH

V/v Thành lập Ban chỉ đạo sản xuất vụ Đông Xuân
Năm 2011 - 2012 trên địa bàn xã Trường Xuân B

ỦY BAN NHÂN DÂN XÃ TRƯỜNG XUÂN B

- Căn cứ Luật tổ chức Hội đồng nhân dân và Ủy Ban nhân dân ngày 26 tháng 11 năm 2003.

- Căn cứ Quyết định số 4987/QĐ-UBND ngày 30/06/2011 của Ủy Ban nhân dân huyện Thới Lai về việc phê chuẩn bầu thành viên Ủy ban nhân dân xã Trường Xuân B, nhiệm kỳ 2011 – 2016;

- Căn cứ vào Kế hoạch số 04/KH-PNN ngày 24/10/2011 của Phòng Nông nghiệp huyện Thới Lai về việc tổ chức thực hiện sản xuất vụ lúa Đông Xuân năm 2011 – 2012;

Xét vào tình hình thực tế trên địa bàn xã Trường Xuân B.

QUYẾT ĐỊNH:

Điều 1 : Nay thành lập Ban chỉ đạo sản xuất vụ Đông Xuân năm 2011-2012 trên địa bàn xã trường Xuân B gồm các Ông (Bà) có danh sách kèm theo.

Điều 2 : Ban chỉ đạo sản xuất vụ Đông Xuân năm 2011-2012 có nhiệm vụ tổ chức tuyên truyền vận động rộng rãi trong nhân dân, nâng cao ý thức về bệnh rầy nâu, bệnh vàng lùn, lùn xoắn lá, làm vệ sinh đồng ruộng trước khi xuống giống vụ Đông Xuân, không sử dụng các loại giống nhiễm rầy, xuống giống đồng loạt đúng lịch thời vụ.

Điều 3 : văn phòng Ủy Ban nhân dân xã, các Ban ngành đoàn thể có liên quan cùng các Ông (Bà) có tên ở điều 1 chịu trách nhiệm thi hành quyết định này.

Nơi nhận :
- Như điều 3;
- Lưu VP.

TM. ỦY BAN NHÂN DÂN XÃ
CHỦ TỊCH

Nguyễn Văn Giữ